Lecture Notes in Computer Science   12450

More information about this series at http://www.springer.com/series/7412

Farah Deeba · Patricia Johnson ·
Tobias Würfl · Jong Chul Ye (Eds.)

Machine Learning for Medical Image Reconstruction

Third International Workshop, MLMIR 2020
Held in Conjunction with MICCAI 2020
Lima, Peru, October 8, 2020
Proceedings

 Springer

Editors
Farah Deeba 🆔
University of British Columbia
Vancouver, BC, Canada

Patricia Johnson 🆔
New York University
New York City, NY, USA

Tobias Würfl 🆔
Friedrich-Alexander University
Erlangen-Nürnberg
Erlangen, Germany

Jong Chul Ye 🆔
Korea Advanced Institute of Science
and Technology
Daejeon, Korea (Republic of)

ISSN 0302-9743   ISSN 1611-3349   (electronic)
Lecture Notes in Computer Science
ISBN 978-3-030-61597-0   ISBN 978-3-030-61598-7   (eBook)
https://doi.org/10.1007/978-3-030-61598-7

LNCS Sublibrary: SL6 – Image Processing, Computer Vision, Pattern Recognition, and Graphics

This Springer imprint is published by the registered company Springer Nature Switzerland AG
The registered company address is: Gewerbestrasse 11, 6330 Cham, Switzerland

Preface

We are proud to present the proceedings for the Third Workshop on Machine Learning for Medical Image Reconstruction (MLMIR 2020) which was held on October 18, 2020, online, as part of the 23rd Medical Image Computing and Computer Assisted Intervention (MICCAI 2020) conference.

Image reconstruction commonly refers to solving an inverse problem, recovering a latent image of some physical parameter from a set of noisy measurements assuming a physical model of the generating process between the image and the measurements. In medical imaging two particular widespread applications are computed tomography (CT) and magnetic resonance imaging (MRI). Using those two modalities as examples, conditions have been established under which the associated reconstruction problems can be solved uniquely. However, in many cases there is a need to recover solutions from fewer measurements to reduce the dose applied to patients or to reduce the measurement time. The theory of compressed sensing showed how to pursue this while still enabling accurate reconstruction by using prior knowledge about the imaged objects. A critical question is the construction of suitable models of prior knowledge about images. In recent years, research has departed from constructing explicit priors for images and moved towards learning suitable priors from large datasets using machine learning (ML).

The aim of this workshop series is to provide a forum for scientific discussion on advanced ML techniques for image acquisition and image reconstruction. After two previous successful workshops, we observe that the interest of the scientific community in this topic has not diminished, and a scientific meeting to foster joint discussions about the topic of image reconstruction is in great demand. Its cross-modality approach brings together researchers from various modalities ranging from CT and MRI to microscopy. We hope that this joint discussion fosters the translation of mathematics and algorithms between those modalities.

We were fortunate that Gitta Kutyniok (LMU Munich, Germany) and Dinggang Shen (United Imaging Intelligence, China) accepted our invitation as keynote speakers and presented fascinating keynote lectures about the state of the art in this emerging field. Despite the special circumstances of the COVID-19 pandemic, we received 19 submissions and accepted 15 papers for inclusion in the workshop. The topics of the accepted papers cover a broad range of medical image reconstruction problems. The predominant ML technique used for reconstruction problems continues to be deep neural networks.

September 2020

Farah Deeba
Patricia Johnson
Tobias Würfl
Jong Chul Ye

Organization

Workshop Organizers

Farah Deeba — University of British Columbia, Canada
Patricia Johnson — New York University, USA
Tobias Würfl — Friedrich-Alexander University Erlangen-Nürnberg, Germany
Jong Chul Ye — Korean Institute of Science and Technology, South Korea

Scientific Program Committee

Jose Caballero — Twitter, USA
Tolga Cukur — Bilkent University, Turkey
Bruno De Man — GE, USA
Enhao Gong — Stanford University, USA
Kerstin Hammernik — Imperial College London, UK
Dong Liang — Chinese Academy of Sciences, China
Ozan Öktem — Royal Institute of Technology, Sweden
Thomas Pock — Graz University of Technology, Austria
Claudia Prieto — King's College London, UK
Essam Rashed — British University in Egypt, Egypt
Matthew Rosen — Havard University, USA
Ge Wang — Rensselaer Polytechnic Institute, USA
Guang Yang — Royal Brompton Hospital, UK

Contents

Deep Learning for General Image Reconstruction

Deep Learning for Magnetic Resonance Imaging

3D FLAT: Feasible Learned Acquisition Trajectories for Accelerated MRI

Jonathan Alush-Aben$^{(\boxtimes)}$, Linor Ackerman-Schraier, Tomer Weiss, Sanketh Vedula, Ortal Senouf, and Alex Bronstein

Technion - Israel Institute of Technology, 32000 Haifa, Israel
jonathandaa@gmail.com

Abstract. Magnetic Resonance Imaging (MRI) has long been considered to be among the gold standards of today's diagnostic imaging. The most significant drawback of MRI is long acquisition times, prohibiting its use in standard practice for some applications. Compressed sensing (CS) proposes to subsample the k-space (the Fourier domain dual to the physical space of spatial coordinates) leading to significantly accelerated acquisition. However, the benefit of compressed sensing has not been fully exploited; most of the sampling densities obtained through CS do not produce a trajectory that obeys the stringent constraints of the MRI machine imposed in practice. Inspired by recent success of deep learning-based approaches for image reconstruction and ideas from computational imaging on learning-based design of imaging systems, we introduce 3D FLAT, a novel protocol for data-driven design of 3D non-Cartesian accelerated trajectories in MRI. Our proposal leverages the entire 3D k-space to simultaneously learn a physically feasible acquisition trajectory with a reconstruction method. Experimental results, performed as a proof-of-concept, suggest that 3D FLAT achieves higher image quality for a given readout time compared to standard trajectories such as radial, stack-of-stars, or 2D learned trajectories (trajectories that evolve only in the 2D plane while fully sampling along the third dimension). Furthermore, we demonstrate evidence supporting the significant benefit of performing MRI acquisitions using non-Cartesian 3D trajectories over 2D non-Cartesian trajectories acquired slice-wise.

Keywords: Magnetic Resonance Imaging · 3D MRI · Fast image acquisition · Image reconstruction · Neural networks · Deep learning · Compressed sensing

1 Introduction

MRI is undoubtedly one of the most powerful tools in use for diagnostic medical imaging due to its noninvasive nature, high resolution, and lack of harmful radiation. It is, however, associated with high costs, driven by relatively expensive hardware and long acquisition times which limit its use in practice. Compressed

© Springer Nature Switzerland AG 2020
F. Deeba et al. (Eds.): MLMIR 2020, LNCS 12450, pp. 3–16, 2020.
https://doi.org/10.1007/978-3-030-61598-7_1

sensing (CS) demonstrated that it is possible to faithfully reconstruct the latent images by observing a fraction of measurements [4]. In [14], the authors demonstrated that it is theoretically possible to accelerate MRI acquisition by randomly sampling the k-space (the frequency domain where the MR images are acquired). However, many CS-based approaches have some practical challenges; it is difficult to construct a feasible trajectory from a given random sampling density or choose k-space frequencies under the constraints.

In addition, the reconstruction of a high-resolution MR image from undersampled measurements is an ill-posed inverse problem where the goal is to estimate the latent image $\mathbf{x}$ (fully-sampled k-space volume) from the observed measurements $\mathbf{y} = \mathcal{F}(\mathbf{x}) + \boldsymbol{\eta}$, where $\mathcal{F}$ is the forward operator (MRI acquisition protocol) and $\boldsymbol{\eta}$ is the sampling noise. Some prior work approached this inverse problem by assuming priors (incorporated in a *maximum a posteriori* setting) on the latent image such as low total variation or sparse representation in a redundant dictionary [14]. Recently, deep supervised learning based approaches have been in the forefront of the MRI reconstruction [8,15,20], solving the above inverse problem through implicitly learning the prior from a data set, and exhibiting significant improvement in the image quality over the explicit prior methods. Other studies, such as SPARKLING [12], have attempted to optimize directly over the feasible k-space trajectories, showing further sizable improvements. The idea of joint optimization of the forward (acquisition) and inverse (reconstruction) processes has been gaining interest in the MRI community for learning sampling patterns [1], Cartesian trajectories [7,19,21] and feasible non-Cartesian 2D trajectories [18].

We distinguish recent works into four paradigms: (i) designing 2D Cartesian trajectories and sampling fully along the third dimension [7,19,21]; (ii) designing 2D sampling densities and performing a full Cartesian sampling along the third dimension [1]; (iii) designing feasible non-Cartesian 2D trajectories and acquiring slice-wise [12,18]; (iv) designing feasible non-Cartesian 3D trajectories, where the design space is unconstrained [13]. This work falls into the final paradigm.

Cartesian sampling limits the degrees of freedom available in k-space data acquisition because it requires sampling fully along one of the dimensions. Acquiring non-Cartesian trajectories in k-space is challenging due to the need of adhering to physical constrains imposed by the machine, namely maximum slew rate of magnetic gradients and upper bounds on the peak currents. [18] developed a method for jointly training 2D image acquisition and reconstruction under these physical constraints, showing promising results and giving inspiration for this work. While there has been an attempt to optimize feasible 3D k-space trajectories in a follow up study on SPARKLING [13], to the best of our knowledge, there has not been any research successful in exploiting the degrees of freedom available in 3D to design sampling trajectories by leveraging the strengths of data-driven reconstruction methods. This is the focus of the present study.

Contributions. We propose 3D Feasible Learned Acquisition Trajectories (3D FLAT), a novel method for data-driven design of 3D non-Cartesian trajectories for MRI; simultaneously optimizing 3D k-space sampling trajectories with an image reconstruction method. We demonstrate that 3D FLAT achieves a significant improvement over standard 3D sampling trajectories - radial and stack-of-stars [11] - under a given time budget. We demonstrate the true merit of performing MRI acquisitions using non-Cartesian 3D trajectories over 2D non-Cartesian trajectories acquired slice-wise. Trajectories learned using 3D FLAT, in some cases, are able to accelerate acquisition by a factor of 2 with no loss in image quality compared to the fixed trajectories using the same reconstruction method.

2   The 3D FLAT Algorithm

Our algorithm can be seen as a pipeline combining the forward (acquisition) and the inverse (reconstruction) models (Fig. 1). The optimization of forward and inverse models is performed simultaneously while imposing physical constraints of the forward model through a penalty term in the loss function. The input to the forward model is the fully sampled k-space, followed by a sub-sampling layer, modeling the data acquisition along a k-space trajectory. The inverse model consists of a re-gridding layer, producing an image on a Cartesian grid in the spatial domain, and a convolutional neural network as a reconstruction model.

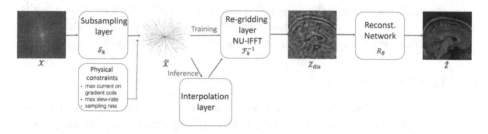

Fig. 1. 3D FLAT data flow pipeline. Notation is explained in the text.

2.1   Forward Model

Sub-sampling Layer. The sub-sampling layer, $\mathcal{S}_{\|}$, emulates MRI acquisition along the trajectory **k**. The trajectory is a tensor **k** of size $N_{shots} \times m \times 3$; N_{shots} is the number of RF (radio frequency) excitations, m is the number of measurements per RF excitation, along three dimensions. The measurements form a complex vector of size $N_{shots} \times m$ emulated by bilinear interpolation $\tilde{\mathbf{x}} = \mathcal{S}_{\mathbf{k}}(\mathbf{X})$ on the full Cartesian grid $\mathbf{X} \in \mathbf{C}^{n \times n \times n}$ where the size of one dimension of the full Cartesian grid is denoted by n, W.L.O.G. assuming the full Cartesian grid is of size $n \times n \times n$. A full Cartesian sampling consists of n^2 RF excitations, each for a line of the volume. We refer to the ratio $\frac{n^2}{N_{shots}}$ as the *acceleration factor* (AF).

In order to obtain an efficient sampling distribution, 3D FLAT uses a coarsened trajectory containing $m' \ll m$ measurements per RF excitation, which is later interpolated to a trajectory of length m with a cubic spline. This approach produces smooth results, well within the physical constraints of the MR machine and allows for efficient training. In addition to this practical advantage, we notice that updating the anchor points and then interpolating encourages more global changes to the trajectory when compared to updating all points. This is partly in spirit with observations made in [3] where linear interpolation is performed after reordering points using a traveling-salesman-problem solver.

2.2 Inverse Model

Regridding Layer. Conventionally, transforming regularly sampled MRI k-space measurements to the image domain requires the inverse fast Fourier transform (IFFT). However, the current case of non-Cartesian sampling trajectories calls for use of the non-uniform inverse FFT (NuFFT) [5], $\hat{\mathcal{F}}_{\mathbf{k}}^{-1}$. The NuFFT performs regridding (resampling and interpolation) of the irregularly sampled points onto a grid followed by IFFT. The result is a (distorted) MR image, $\mathbf{Z}_{dis} = \hat{\mathcal{F}}_{\mathbf{k}}^{-1}(\tilde{\mathbf{x}})$.

Reconstruction Model. The reconstruction model extracts the latent image $\hat{\mathbf{Z}}$ from the distorted image $\mathbf{Z}_{dis}$; $\hat{\mathbf{Z}} = R_{\theta}(\mathbf{Z}_{dis})$, R represents the model and θ its learnable parameters. The reconstruction model passes the gradients back to the forward model in order to update the trajectory $\mathbf{k}$ so it will contribute most to the reconstruction quality. We emphasize that the principal focus of this work is not on the reconstruction model itself, and the proposed algorithm can be used with any differentiable model to improve the end-task performance.

2.3 Loss Function

The pipeline is trained by simultaneously learning the trajectory $\mathbf{k}$ and the parameters of the reconstruction model θ. To optimize the reconstruction performance while maintaining a feasible trajectory, we used a loss function composed of a data fidelity term and a constraint violation term, $L = L_{task} + L_{const}$.

Data Fidelity. The L_1 norm is used to measure the discrepancy between the model output image $\hat{\mathbf{Z}}$ and the ground-truth image $\mathbf{Z} = \mathcal{F}^{-1}(\mathbf{X})$, derived from the fully sampled k-space, $L_{task} = \|\hat{\mathbf{Z}} - \mathcal{F}^{-1}(\mathbf{X})\|_1$.

Machine Constraints. A feasible sampling trajectory must follow the physical hardware constraints of the MRI machine, specifically the peak-current (translated into the maximum value of imaging gradients $G_{\max}$), along with the maximum slew-rate $S_{\max}$ produced by the gradient coils. These requirements can be translated into geometric constraints on the first and second-order derivatives of each of the spatial coordinates of the trajectory: $|\dot{k}| \approx \frac{|k_{i+1}-k_i|}{dt} \leq v_{\max} = \gamma\, G_{\max}$ and $|\ddot{k}| \approx \frac{|k_{i+1}-2k_i+k_{i-1}|}{dt^2} \leq a_{\max} = \gamma\, S_{\max}$ (γ is the gyromagnetic ratio).

The constraint violation term L_{const} in the loss function applies to the trajectory $\mathbf{k}$ only and penalizes it for violation of the physical constraints. We chose the hinge functions of the form $\max(0, |\dot{k}| - v_{\max})$ and $\max(0, |\ddot{k}| - a_{\max})$ summed over the trajectory spatial coordinates and over all sample points. These penalties remain zero as long as the solution is feasible and grow linearly with the violation of each of the constraints. The relative importance of the velocity (peak current) and acceleration (slew rate) penalties is governed by the parameters λ_v and λ_a, respectively. Note that in case of learning 3D trajectories, the constraints are enforced in 3 dimensions, corresponding to the respective gradient coils.

Optimization. The training is carried out by solving the optimization problem

$$\min_{\mathbf{k},\theta} \sum_{(\mathbf{X},\mathbf{Z})} L_{task}(R_\theta(\hat{\mathcal{F}}_{\mathbf{k}}^{-1}(\mathcal{S}_{\mathbf{k}}(\mathbf{X}))), \mathbf{Z}) + L_{const}(\mathbf{k}), \tag{1}$$

where the loss is summed over a training set comprising the pairs of fully sampled data $\mathbf{X}$ and the corresponding groundtruth output $\mathbf{Z}$.

3   Experimental Evaluation[1]

3.1   Dataset

T1-weighted images taken from the human connectome project (HCP) [17] were used. We down-sampled the HCP's 1065 brain MRI volumes to $80 \times 80 \times 80$, from the original $145 \times 174 \times 145$, keeping an isotropic spatial resolution, and using a 90/10 split for training/validation.

3.2   Training Settings

The network was trained using the Adam [10] optimizer. Learning rate was set to 0.001 for the reconstruction model, and 0.005 for the sub-sampling layer. For the differentiable regridding layer (Nu-IFFT) [5], we made use of an initial 2D implementation available from the authors of [18][2]. For the reconstruction model, we used a 3D U-Net architecture [23], based on the publicly-available implementation[3]. U-Net is widely-used in medical imaging tasks in general, and in MRI reconstruction [20] and segmentation [9] in particular. We emphasize that the scope of this work is not directed toward building the best reconstruction method, but rather demonstrating the benefit of simultaneous optimization of the acquisition-reconstruction pipeline; any differentiable reconstruction method is suitable. The physical constraints we enforced are: $G_{\max} = 40\,\mathrm{mT/m}$ for the peak gradient, $S_{\max} = 200\,\mathrm{T/m/s}$ for the maximum slew-rate, and $dt = 10\,\mu\mathrm{sec}$ for the sampling time.

3.3   Reference Trajectories

Standard trajectories used in 3D multi-shot MR imaging are radial lines and stacks of stars (SOS) [6]. SOS is a 2D radial trajectory in the xy plane multiplexed over the z dimension. In the experiment *SOS 2D*, a trajectory was initialized with SOS and learned in the xy plane only. We claim 3D-FLAT can

---

[1] Our code is available at https://github.com/3d-flat/3dflat.
[2] https://github.com/tomer196/PILOT.
[3] https://github.com/wolny/pytorch-3dunet.

optimize any heuristically hand-crafted trajectory, but limit our choice to these radial initializations. Other trajectories such as Wave-CAIPI as proposed in [2], could be used as well. All slices were acquired with the same trajectory as in the multi-shot PILOT experiment suggested in [18]. In the experiment *SOS 3D*, a trajectory initialized with SOS was allowed to train in 3D, exploring all degrees of freedom available. The 3D radial trajectories were constructed as described in [11], using the MATLAB implementation[4]. The 2D radial trajectories were evenly distributed around the center. In our simulation, sampling $m = 3000$ data-points over per shot of N_{shots} did not add any new information, unlike a real sampling scenario. For ease of computation, $m' = 100$ points were sampled.

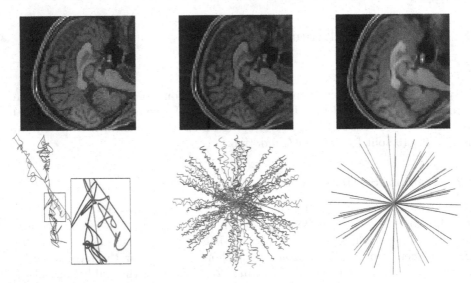

Fig. 2. First row: reconstruction results using different sampling methods (left-to-right): groundtruth image using full k-space; 3D FLAT with radial initialization; and fixed radial trajectory. Second row: depiction of the trajectories (left-to-right): 2 shots of 3D FLAT learned trajectory and its initialization overlaid; 3D FLAT learned trajectory; and radial trajectory used for initializing for 3D FLAT. Note that the second row presents 3D trajectories but visualized in 2D.

3.4   Results and Discussion

For quantative evaluation, we use the peak signal-to-noise ratio (PSNR) and structural-similarity (SSIM) [22] measures. All trajectories used in our experiments are feasible, satisfying mentioned machine constraints. We compared our algorithms to training a reconstruction model using measurements obtained using fixed handcrafted trajectories, "fixed trajectories". Quantitative results (Fig. 3) show that every 3D FLAT trajectory (the learned SOS 2D/3D and radial) outperforms fixed trajectories in every acceleration factor (AF). We notice an improvement of $1.01 - 3.4$ dB in PSNR and $0.0452 - 0.0896$ in SSIM on full 3D trajectories.

---

[4] https://github.com/LarsonLab/Radial-Field-of-Views.

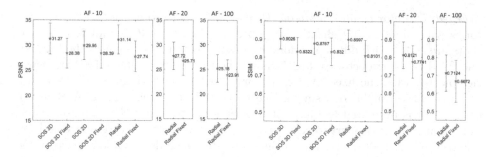

Fig. 3. Quantitative results (PSNR & SSIM) of 3D FLAT for different acceleration factors and initializations. 3D FLAT outperforms fixed trajectories in all acceleration factors and initialization. Error bars report best, worst and average values seen.

The standard deviation over the SSIM metric is noticeably smaller in some cases of 3D FLAT trajectories. Notice in Fig. 3, achieved mean PSNR of the learned radial trajectory with AF of 20 performs similarly to that of a fixed radial trajectory with AF of 10, yielding a speedup factor of 2 with no loss in image quality. Furthermore, results suggest that learning the trajectory in all three dimensions (learned SOS 3D and learned radial) leads to a more significant improvement (1.32 dB in PSNR) over the same initialization allowed to learn in two dimensions. This corroborates the natural assumption that an additional degree of freedom (3D) used to design trajectories can improve the end image quality. This improves the result of 3D SPARKLING [13] where, to the best of our understanding, the authors reached a conclusion that 2D SPARKLING trajectories acquired slice-wise outperformed the 3D ones. The possible limitations of 3D SPARKLING which 3D FLAT alleviates are twofold: Firstly, 3D SPARKLING enforces constraints on the search space of feasible trajectories by replicating the same learned trajectory in multiple regions of the k-space whereas in 3D FLAT the search space is unconstrained. Secondly, 3D SPARKLING requires an estimate of the desired sampling density prior to optimization whereas 3D FLAT enables task-driven learning of optimal sampling density jointly with the feasible trajectories (Fig. 4).

Qualitative results in Figs. 6, 8 and 9 present visual depiction of sample slices from multiple views obtained using different acquisition trajectories (learned and fixed) at different AFs. The visual results suggest that learned trajectories contain more details and are of superior quality than the fixed counterparts. That said, performing experiments on real machines is necessary for the next steps of the research, but out of the scope of this proof-of-concept.

We tested our algorithm across three AFs, 10, 20, and 100, where they demonstrated invariable success in improving the reconstruction accuracy. We notice the radial trajectory with AF 20 performs as well as the fixed radial trajectory of AF 10. To demonstrate the robustness of 3D FLAT to different reconstruction methods, we performed reconstructions using off-the-shelf TV-regularized compressed-sensing inverse problem solvers [16]. The results are presented in Table 1 in the Appendix comparing 3D FLAT with fixed counterparts. In all cases, 3D FLAT outperformed fixed trajectories in terms of PSNR.

Learned Trajectories and Sampling Densities. Visualizations of learned trajectories with different initializations are presented in Figs. 2, 5, and 7, Different shots are coded with different colors. Figure 2 (bottom row) depicts the learned and fixed radial trajectories, Fig. 5 shows the learned SOS 2D and 3D trajectories plotted with the fixed SOS trajectory. Note that the learned SOS 2D and 3D trajectories might look similar due to visualization limitations. A close up of one shot is presented in Fig. 7; it is interesting to notice the increased sampling density in high-curvature regions of the trajectory. The reason could be due to the enforced constraints on the slew-rate which do not allow sharp turns in the trajectory, resulting in increased sampling in these regions. Furthermore, a visualization of sampling density of learned and fixed trajectories is presented in Fig. 4. We see the learned radial and SOS 3D have improved sampling density at the center of the k-space. This coincides with the intuition suggesting that the center containing more low-frequency information is more important for reconstruction. This is a possible reason for 3D FLAT outperforming the fixed trajectories while using TV-regularized direct reconstruction. We also notice a dependence of the learned trajectories on initialization, this problem was also encountered by [12] and [18], this could be due to local minima near them.

4   Conclusion

We demonstrated, as a proof-of-concept, that learning-based design of feasible non-Cartesian 3D trajectories in MR imaging leads to better image reconstruction when compared to the off-the-shelf trajectories. To the best of our knowledge, this is the first attempt of data-driven design of feasible 3D trajectories in MRI. We further demonstrate the benefit of acquiring 3D non-Cartesian trajectories over their 2D counterparts acquired slice-wise. Our experiments suggest that the learned trajectories fall significantly below the enforced machine constraints. We believe that such trajectories can be deployed in low magnetic field portable MRI scanners (i.e. Hyperfine[5]) to achieve better reconstruction accuracy vs. acquisition speed. We plan to try this in future work. We defer the following aspects to future work: Firstly, in this work, we limited our attention to a relatively small resolution of the k-space. The reason for this is due to the computational complexity of our reconstruction method which can be alleviated by using 3D CNNs with lower complexity. Secondly, this work did not take into account signal decay within a single acquisition. This noise can be modelled and taken into account during training. However, since each shot is relatively small (30 ms), we believe there would not be a noticeable effect on the final reconstruction. Thirdly, the trajectories achieved are sub-optimal, as they are highly dependant on initialization. This could be improved by optimizing in a two step process: optimizing for the optimal sampling density and then designing a trajectory while enforcing machine constraints, this was proposed in [18] in a 2D single-shot scenario (PILOT-TSP). Lastly, this work is a proof-of-concept that has been validated through simulations and has yet to be validated on real MRI machines.

---

[5] https://www.hyperfine.io/.

A   Supplementary Material

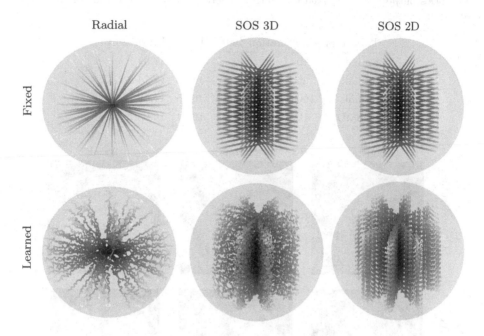

Fig. 4. The sampling densities of the fixed trajectories and 3D FLAT are visualized above. Notice the change in density between the fixed initialization and the learned. Note that for visualization purposes only a fraction of the points are shown. Best viewed in color.

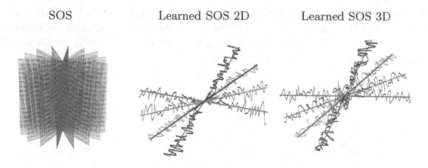

Fig. 5. Compared above are three trajectories initialized using stack-of-stars. From left to right are: fixed trajectory; 4 shots of learned SOS 2D trajectories; 4 corresponding shots of learned SOS 3D trajectories. Note that the right most trajectory is actually 3D and that all the trajectories obey the MR physical constraints.

Table 1. Comparison of 3D FLAT and fixed trajectories with TV-regularized image reconstruction using off-the-shelf CS inverse problem solvers (BART, [16]). Learned trajectories outperform the fixed counterparts across all acceleration factors and initializations.

Trajectory	Acceleration factor	Fixed	Learned
Radial	10	17.19	17.99
Radial	20	16.98	17.24
Radial	100	16.75	17.5
SOS 3D	10	16.14	16.98

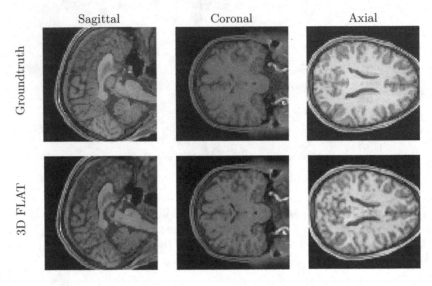

Fig. 6. The space is processed in 3D, any plane can be depicted easily. Shown above are three planes of the same volume. The first row is the ground truth image, processed with the full k-space. The second was created with 3D FLAT initialized with a radial trajectory at an acceleration factor of 10.

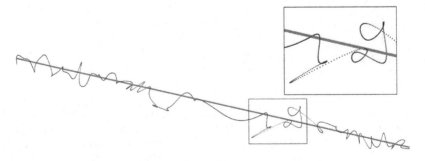

Fig. 7. A single radial trajectory is shown with its initialization. Notice the density at the high curvature parts of the black curve.

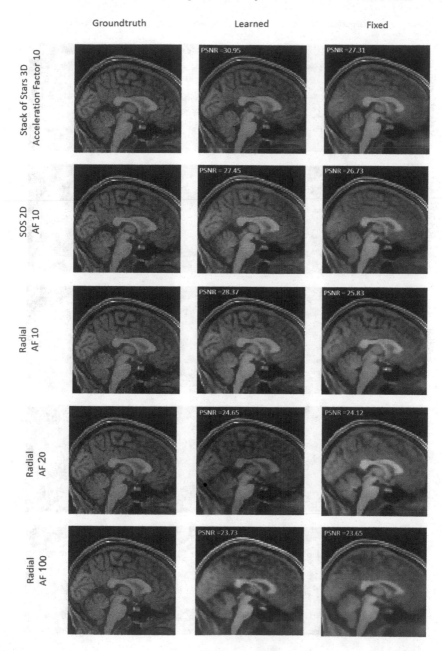

Fig. 8. Depicted are reconstruction results of all trajectories over different acceleration factors. The images depict a sagittal plane of a sample volume. PSNR is calculated w.r.to the groundtruth image on the left most column.

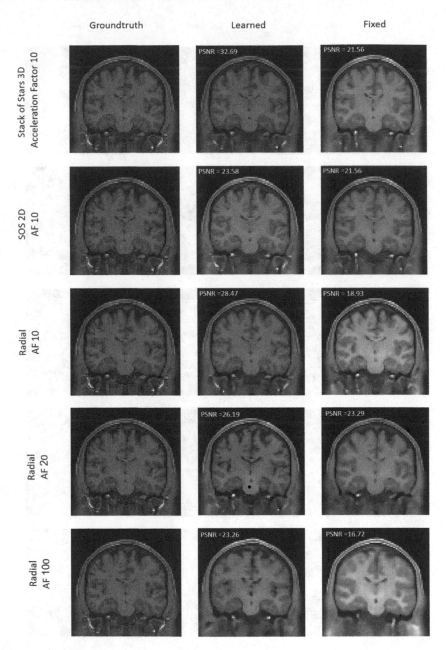

Fig. 9. Depicted are reconstruction results of all trajectories over different acceleration factors. The images depict a coronal plane of a sample volume. PSNR is calculated w.r.to the groundtruth image on the left most column.

References

1. Bahadir, C.D., Dalca, A.V., Sabuncu, M.R.: Learning-based optimization of the under-sampling pattern in MRI. In: Chung, A.C.S., Gee, J.C., Yushkevich, P.A., Bao, S. (eds.) IPMI 2019. LNCS, vol. 11492, pp. 780–792. Springer, Cham (2019). https://doi.org/10.1007/978-3-030-20351-1_61
2. Bilgic, B., et al.: Wave-CAIPI for highly accelerated 3D imaging. Magn. Reson. Med. **73**(6), 2152–2162 (2015). https://doi.org/10.1002/mrm.25347. https://onlinelibrary.wiley.com/doi/abs/10.1002/mrm.25347
3. Boyer, C., Chauffert, N., Ciuciu, P., Kahn, J., Weiss, P.: On the generation of sampling schemes for magnetic resonance imaging. SIAM J. Imaging Sci. **9**, 2039–2072 (2016)
4. Candès, E.J., Romberg, J., Tao, T.: Robust uncertainty principles: exact signal reconstruction from highly incomplete frequency information. IEEE Trans. Inf. Theory **52**, 489–509 (2006)
5. Dutt, A., Rokhlin, V.: Fast Fourier transforms for nonequispaced data. SIAM J. Sci. Comput. **14**, 1368–1393 (1993)
6. Glover, G.H., Pauly, J.M.: Projection reconstruction techniques for reduction of motion effects in MRI. Magn. Reson. Med. **28**, 275–289 (1992)
7. Gözcü, B., et al.: Learning-based compressive MRI. IEEE Trans. Med. Imaging **37**, 1394–1406 (2018)
8. Hammernik, K., et al.: Learning a variational network for reconstruction of accelerated MRI data. Magn. Reson. Med. **79**, 3055–3071 (2018)
9. Isensee, F., Jaeger, P.F., Full, P.M., Wolf, I., Engelhardt, S., Maier-Hein, K.H.: Automatic cardiac disease assessment on cine-MRI via time-series segmentation and domain specific features. In: Pop, M., et al. (eds.) STACOM 2017. LNCS, vol. 10663, pp. 120–129. Springer, Cham (2018). https://doi.org/10.1007/978-3-319-75541-0_13
10. Kingma, D.P., Ba, J.: Adam: a method for stochastic optimization. arXiv e-prints, December 2014
11. Larson, P., Gurney, P., Nishimura, D.: Anisotropic field-of-views in radial imaging. IEEE Trans. Med. Imaging **27**, 47–57 (2007)
12. Lazarus, C., et al.: SPARKLING: variable-density k-space filling curves for accelerated T2*-weighted MRI. Magn. Reson. Med. **81**, 3643–3661 (2019)
13. Lazarus, C., Weiss, P., Gueddari, L., Mauconduit, F., Vignaud, A., Ciuciu, P.: 3D sparkling trajectories for high-resolution T2*-weighted magnetic resonance imaging (2019)
14. Lustig, M., Donoho, D., Pauly, J.M.: Sparse MRI: the application of compressed sensing for rapid MR imaging. Magn. Reson. Med. Official J. Int. Soc. Magn. Reson. Med. **58**, 1182–1195 (2007)
15. Sun, J., Li, H., Xu, Z., et al.: Deep ADMM-Net for compressive sensing MRI. In: Advances in Neural Information Processing Systems (2016)
16. Uecker, M., et al.: ESPIRiT—an eigenvalue approach to autocalibrating parallel MRI: where SENSE meets GRAPPA. Magn. Reson. Med. Official J. Int. Soc. Magn. Reson. Med. **71**, 990–1001 (2014)
17. Van Essen, D., Ugurbil, K., et al.: The human connectome project: a data acquisition perspective. NeuroImage (2012). https://doi.org/10.1016/j.neuroimage.2012.02.018. Connectivity
18. Weiss, T., Senouf, O., Vedula, S., Michailovich, O., Zibulevsky, M., Bronstein, A.: PILOT: physics-informed learned optimal trajectories for accelerated MRI. arXiv e-prints, September 2019

19. Weiss, T., Vedula, S., Senouf, O., Bronstein, A., Michailovich, O., Zibulevsky, M.: Learning fast magnetic resonance imaging. arXiv e-prints arXiv:1905.09324, May 2019
20. Zbontar, J., et al.: fastMRI: an open dataset and benchmarks for accelerated MRI. arXiv preprint arXiv:1811.08839 (2018)
21. Zhang, Z., Romero, A., Muckley, M.J., Vincent, P., Yang, L., Drozdzal, M.: Reducing uncertainty in undersampled MRI reconstruction with active acquisition. arXiv preprint arXiv:1902.03051 (2019)
22. Wang, Z., Bovik, A.C., Sheikh, H.R., Simoncelli, E.P.: Image quality assessment: from error visibility to structural similarity. IEEE Trans. Image Processing **13**, 600–612 (2004)
23. Çiçek, O., Abdulkadir, A., Lienkamp, S., Brox, T., Ronneberger, O.: 3D U-Net: learning dense volumetric segmentation from sparse annotation. arXiv arXiv:1606.06650 (2016)

Deep Parallel MRI Reconstruction
Network Without Coil Sensitivities

Wanyu Bian[1], Yunmei Chen[1(✉)], and Xiaojing Ye[2]

[1] University of Florida, Gainesville, FL 32611, USA
{wanyu.bian,yun}@ufl.edu
[2] Georgia State University, Atlanta, GA 30302, USA
xye@gsu.edu

Abstract. We propose a novel deep neural network architecture by mapping the robust proximal gradient scheme for fast image reconstruction in parallel MRI (pMRI) with regularization function trained from data. The proposed network learns to adaptively combine the multi-coil images from incomplete pMRI data into a single image with homogeneous contrast, which is then passed to a nonlinear encoder to efficiently extract sparse features of the image. Unlike most of existing deep image reconstruction networks, our network does not require knowledge of sensitivity maps, which can be difficult to estimate accurately, and have been a major bottleneck of image reconstruction in real-world pMRI applications. The experimental results demonstrate the promising performance of our method on a variety of pMRI imaging data sets.

Keywords: Proximal gradient · Parallel MRI · Coil sensitivity

1 Introduction

Parallel magnetic resonance imaging (pMRI) is a state-of-the-art medical MR imaging technology which surround the scanned objects by multiple receiver coils and collect k-space (Fourier) data in parallel. To accelerate scan process, partial data acquisitions that increase the spacing between read-out lines in k-space are implemented in pMRI. However, reduction in k-space data sampling arising aliasing artifacts in images, which must be removed by image reconstruction process. There are two major approaches to image reconstruction in pMRI: the first approach are k-space methods which interpolate the non-sampled k-space data using the sampled ones across multiple receiver coils [3], such as the generalized auto-calibrating partially parallel acquisition (GRAPPA) [6]. The other approach is the class of image space methods which remove the aliasing artifacts in the image domain by solving a system of equations that relate the image to

Electronic supplementary material The online version of this chapter (https:// doi.org/10.1007/978-3-030-61598-7_2) contains supplementary material, which is available to authorized users.

F. Deeba et al. (Eds.): MLMIR 2020, LNCS 12450, pp. 17–26, 2020.
https://doi.org/10.1007/978-3-030-61598-7_2

be reconstructed and partial k-spaced data through coil sensitivities, such as in SENSitivity Encoding (SENSE) [13].

In this paper, we propose a new deep learning based reconstruction method to address several critical issues of pMRI reconstruction in image space. Consider a pMRI system with N_c receiver coils acquiring 2D MR images at resolution $m \times n$ (we treat a 2D image $\mathbf{v} \in \mathbb{C}^{m \times n}$ and its column vector form $\mathbf{v} \in \mathbb{C}^{mn}$ interchangeably hereafter). Let $\mathbf{P} \in \mathbb{R}^{p \times mn}$ be the binary matrix representing the undersampling mask with p sample locations in k-space, and $\mathbf{s}_i \in \mathbb{C}^{mn}$ the coil sensitivity and $\mathbf{f}_i \in \mathbb{C}^p$ the *partial* k-space data at the ith receiver coil for $i = 1, \ldots, N_c$. Therefore $\mathbf{f}_i$ and the image $\mathbf{v}$ are related by $\mathbf{f}_i = \mathbf{PF}(\mathbf{s}_i \cdot \mathbf{v}) + \mathbf{n}_i$ where $\cdot$ denotes pointwise multiplication of two matrices, and $\mathbf{n}_i$ is the unknown acquisition noise in k-space at each receiver coil. Then SENSE-based image space reconstruction methods can be generally formulated as an optimization problem:

$$\min_{\mathbf{v}} \sum_{i=1}^{N_c} \frac{1}{2} \|\mathbf{PF}(\mathbf{s}_i \cdot \mathbf{v}) - \mathbf{f}_i\|^2 + R(\mathbf{v}), \tag{1}$$

where $\mathbf{v} \in \mathbb{C}^{mn}$ is the MR image to be reconstructed, $\mathbf{F} \in \mathbb{C}^{mn \times mn}$ stands for the discrete Fourier transform, and $R(\mathbf{v})$ is the regularization on the image $\mathbf{v}$. $\|\mathbf{x}\|^2 := \|\mathbf{x}\|_2^2 = \sum_{j=1}^n |x_j|^2$ for any complex vector $\mathbf{x} = (x_1, \ldots, x_n)^\top \in \mathbb{C}^n$. There are two critical issues in pMRI image reconstruction using (1): availability of accurate coil sensitivities $\{\mathbf{s}_i\}$ and proper image regularization R. Most existing SENSE-based reconstruction methods assume coil sensitivity maps are given, which are however difficult to estimate accurately in real-world applications. On the other hand, the regularization R is of paramount importance to the inverse problem (1) to produce desired images from significantly undersampled data, but a large number of existing methods employ handcrafted regularization which are incapable to extract complex features from images effectively.

In this paper, we tackle the two aforementioned issues in an unified deep-learning framework dubbed as pMRI-Net. Specifically, we consider the reconstruction of multi-coil images $\mathbf{u} = (\mathbf{u}_1, \ldots, \mathbf{u}_{N_c}) \in \mathbb{C}^{mn N_c}$ for all receiver coils to avoid use of coil sensitivity maps (but can recover them as a byproduct), and design a deep residual network which can jointly learn the adaptive combination of multi-coil images and an effective regularization from training data.

The contribution of this paper could be summarized as follows: Our method is the first "combine-then-regularize" approach for deep-learning based pMRI image reconstruction. The combination operator integrates multichannel images into single channel and this approach performs better than the linear combination the root of sum-of-squares (SOS) method [13]. This approach has three main advantages: (i) the combined image has homogeneous contrast across the FOV, which makes it suitable for feature-based image regularization and less affected by the intensity biases in coil images; (ii) the regularization operators are applied to this single body image in each iteration, and require much fewer network parameters to reduce overfitting and improve robustness; and (iii) our approach naturally avoids the use of sensitivity maps, which has been a thorny issue in image-based pMRI reconstruction.

2   Related Work

Most existing deep-learning (DL) based methods rendering end-to-end neural networks mapping from the partial k-space data to the reconstructed images [9,11,14,19,20]. The common issue with this class of methods is that the DNNs require excessive amount of data to train, and the resulting networks perform similar to "black-boxes" which are difficult to interpret and modify.

In recent years, a class of DL based methods improve over the end-to-end training by selecting the scheme of an iterative optimization algorithm and pre-scribe a phase number T, map each iteration of the scheme to one phase of the network. These methods are often known as the learned optimization algorithms (LOAs) [1,2,7,15,22–24]. For instance, ADMM-Net [23], ISTA-Net$^+$ [24], and cascade network [15] are regular MRI reconstruction. For pMRI: Variational network (VN) [7] introduced gradient descent method by applying given sensitivities $\{\mathbf{s}_i\}$. MoDL [1] proposed a recursive network by unrolling the conjugate gradient algorithm using a weight sharing strategy. Blind-PMRI-Net [12] designed three network blocks to alternately update multi-channel images, sensitivity maps and the reconstructed MR image using an iterative algorithm based on half-quadratic splitting. The network in [16] developed a Bayesian framework for joint MRI-PET reconstruction. VS-Net [4] derived a variable splitting optimization method. However, existing methods still face the lack of accurate coil sensitivity maps and proper regularization in the pMRI problem.

Recently, a method called DeepcomplexMRI [20] developed an end-to-end learning without explicitly using coil sensitivity maps to recover channel-wise images, and then combine to a single channel image in testing.

This paper proposes a novel deep neural network architecture which integrating the robust proximal gradient scheme for pMRI reconstruction without knowledge of coil sensitivity maps. Our network learns to adaptively combine the channel-wise image from the incomplete data to assist the reconstruction and learn a nonlinear mapping to efficiently extract sparse features of the image by using a set of training data on the pairs of under-sampled channel-wise k-space data and corresponding images. The roles of the multi-coil image combination operator and sparse feature encoder are clearly defined and jointly learned in each iteration. As a result, our network is more data efficient in training and the reconstruction results are more accurate.

3   Proposed Method

3.1   Joint Image Reconstruction pMRI Without Coil Sensitivities

We propose an alternative pMRI reconstruction approach to (1) by recovering images from individual receiver coils jointly. Denote $\mathbf{u}_i$ the MR image at the ith receiver coil, i.e., $\mathbf{u}_i = \mathbf{s}_i \cdot \mathbf{v}$, where the sensitivity $\mathbf{s}_i$ and the full FOV image $\mathbf{v}$ are both unknown in practice. Thus, the image $\mathbf{u}_i$ relates to the partial k-space data $\mathbf{f}_i$ by $\mathbf{f}_i = \mathbf{PFu}_i + \mathbf{n}_i$, and hence the data fidelity term is formulated as least squares $(1/2) \cdot \|\mathbf{PFu}_i - \mathbf{f}_i\|^2$. We also need a suitable regularization R on the images $\{\mathbf{u}_i\}$. However, these images have vastly different contrasts due to the significant variations in the sensitivity maps at different receiver coils. Therefore, it is more appropriate to apply regularization to the (unknown) image $\mathbf{v}$.

To address the issue of regularization, we propose to first learn a nonlinear operator $\mathcal{J}$ that combines $\{\mathbf{u}_i\}$ into the image $\mathbf{v} = \mathcal{J}(\mathbf{u}_1, \ldots, \mathbf{u}_{N_c}) \in \mathbb{C}^{m \times n}$ with homogeneous contrast, and apply a regularization on $\mathbf{v}$ with a parametric form $\|\mathcal{G}(\mathbf{v})\|_{2,1}$ by leveraging the robust sparse selection property of $\ell_{2,1}$-norm and shrinkage threshold operator. Here $\mathcal{G}(\mathbf{v})$ represents a nonlinear sparse encoder trained from data to effectively extract complex features from the image $\mathbf{v}$. Combined with the data fidelity term above, we propose the following pMRI image reconstruction model:

$$\mathbf{u}(\mathbf{f}; \Theta) = \arg\min_{\mathbf{u}} \frac{1}{2} \sum_{i=1}^{N_c} \|\mathbf{PFu}_i - \mathbf{f}_i\|_2^2 + \|\mathcal{G} \circ \mathcal{J}(\mathbf{u})\|_{2,1}, \tag{2}$$

where $\mathbf{u} = (\mathbf{u}_1, \ldots, \mathbf{u}_{N_c})$ is the multi-channel image to be reconstructed from the pMRI data $\mathbf{f} = (\mathbf{f}_1, \ldots, \mathbf{f}_{N_c})$, and $\Theta = (\mathcal{G}, \mathcal{J})$ represents the parameters of the deep networks $\mathcal{G}$ and $\mathcal{J}$. The key ingredients of (2) are the nonlinear combination operator $\mathcal{J}$ and sparse feature encoder $\mathcal{G}$, which we describe in details in Sect. 3.3. Given a training data set consisting of J pairs $\{(\mathbf{f}^{[j]}, \hat{\mathbf{u}}^{[j]}) \mid 1 \le j \le J\}$, where $\mathbf{f}^{[j]} = (\mathbf{f}_1^{[j]}, \ldots, \mathbf{f}_{N_c}^{[j]})$ and $\hat{\mathbf{u}}^{[j]} = (\hat{\mathbf{u}}_1^{[j]}, \ldots, \hat{\mathbf{u}}_{N_c}^{[j]})$ are respectively the partial k-space data and the ground truth image reconstructed by full k-space data of the jth image data, our goal is to learn Θ (i.e., $\mathcal{G}$ and $\mathcal{J}$) from the following bi-level optimization problem:

$$\min_{\Theta} \frac{1}{J} \sum_{j=1}^{J} \ell(\mathbf{u}(\mathbf{f}^{[j]}; \Theta), \hat{\mathbf{u}}^{[j]}), \text{ s.t. } \mathbf{u}(\mathbf{f}^{[j]}; \Theta) \text{ solves } (2) \text{ with data } \mathbf{f}^{[j]}, \tag{3}$$

where $\ell(\mathbf{u}, \hat{\mathbf{u}})$ measures the discrepancy between the reconstruction $\mathbf{u}$ and the ground truth $\hat{\mathbf{u}}$. To tackle the lower-level minimization problem in (3), we construct a proximal gradient network with residual learning as an (approximate) solver of (2). Details on the derivation of this network are provided in the next subsection.

3.2 Proximal Gradient Network with Residual Learning

If the operators $\mathcal{J}$ and $\mathcal{G}$ were given, we can apply proximal gradient descent algorithm to approximate a (local) minimizer of (2) by iterating

$$\mathbf{b}_i^{(t)} = \mathbf{u}_i^{(t)} - \rho_t \mathbf{F}^\top \mathbf{P}^\top (\mathbf{PFu}_i^{(t)} - \mathbf{f}_i), \tag{4a}$$

$$\mathbf{u}_i^{(t+1)} = [\text{prox}_{\rho_t \|\mathcal{G} \circ \mathcal{J}(\cdot)\|_{2,1}}(\mathbf{b}^{(t)})]_i, \quad 1 \le i \le N_c \tag{4b}$$

where $\mathbf{b}^{(t)} = (\mathbf{b}_1^{(t)}, \ldots, \mathbf{b}_{N_c}^{(t)})$, $[\mathbf{x}]_i = \mathbf{x}_i \in \mathbb{C}^{mn}$ for any vector $\mathbf{x} \in \mathbb{C}^{mnN_c}$, $\rho_t > 0$ is the step size, and prox_g is the proximal operator of g defined by

$$\text{prox}_g(\mathbf{b}) = \arg\min_{\mathbf{x}} g(\mathbf{x}) + \frac{1}{2}\|\mathbf{x} - \mathbf{b}\|^2. \tag{5}$$

The gradient update step (4a) is straightforward to compute and fully utilizes the relation between the partial k-space data $\mathbf{f}_i$ and the image $\mathbf{u}_i$ to be reconstructed as derived from MRI physics. The proximal update step (4b), however, presents several difficulties: the operators $\mathcal{J}$ and $\mathcal{G}$ are unknown and need to be learned from data, and the proximal operator $\mathrm{prox}_{\rho_t \|\mathcal{G} \circ \mathcal{J}(\cdot)\|_{2,1}}$ most likely will not have closed form and can be difficult to compute. Assuming that we have both $\mathcal{J}$ and $\mathcal{G}$ parametrized by convolutional networks, we adopt a residual learning technique by leveraging the shrinkage operator (as the proximal operator of $\ell_{2,1}$-norm $\|\cdot\|_{2,1}$) and converting (4b) into an explicit update formula. To this end, we parametrize the proximal step (4b) as an implicit residual update:

$$\mathbf{u}_i^{(t+1)} = \mathbf{b}_i^{(t)} + [\mathbf{r}(\mathbf{u}_1^{(t+1)}, \cdots, \mathbf{u}_{N_c}^{(t+1)})]_i, \tag{6}$$

where $\mathbf{r} = \tilde{\mathcal{J}} \circ \tilde{\mathcal{G}} \circ \mathcal{G} \circ \mathcal{J}$ is the residual network as the composition of $\mathcal{J}$, $\mathcal{G}$, and their adjoint operators $\tilde{\mathcal{J}}$ and $\tilde{\mathcal{G}}$. These four operators are learned separately to increase the capacity of the network. To reveal the role of nonlinear shrinkage selection in (6), consider the original proximal update (4b) where

$$\mathbf{u}^{(t+1)} = \arg\min_{\mathbf{u}} \|\mathcal{G} \circ \mathcal{J}(\mathbf{u})\|_{2,1} + \frac{1}{2\rho_t} \|\mathbf{u} - \mathbf{b}^{(t)}\|^2. \tag{7}$$

For certain convolutional networks $\mathcal{J}$ and $\mathcal{G}$ with rectified linear unit (ReLU) activation, $\|\mathbf{u} - \mathbf{b}^{(t)}\|^2$ can be approximated by $\alpha \|\mathcal{G} \circ \mathcal{J}(\mathbf{u}) - \mathcal{G} \circ \mathcal{J}(\mathbf{b}^{(t)})\|^2$ for some $\alpha > 0$ dependent on $\mathcal{J}$ and $\mathcal{G}$ [24]. Substituting this approximation into (7), we obtain that

$$\mathcal{G} \circ \mathcal{J}(\mathbf{u}^{(t+1)}) = \mathcal{S}_{\alpha_t}(\mathcal{G} \circ \mathcal{J}(\mathbf{b}^{(t)})), \tag{8}$$

where $\alpha_t = \rho_t/\alpha$, $\mathcal{S}_{\alpha_k}(\mathbf{x}) = \mathrm{prox}_{\alpha_k \|\cdot\|_{2,1}}(\mathbf{x}) = [\mathrm{sign}(x_i) \max(|x_i| - \alpha_k, 0)] \in \mathbb{R}^n$ for any vector $\mathbf{x} = (x_1, \ldots, x_n) \in \mathbb{R}^n$ is the soft shrinkage operator. Plugging (8) into (6), we obtain an explicit form of (4b), which we summarize together with (4a) in the following scheme:

$$\mathbf{b}_i^{(t)} = \mathbf{u}_i^{(t)} - \rho_t \mathbf{F}^\top \mathbf{P}^\top (\mathbf{PF}\mathbf{u}_i^{(t)} - \mathbf{f}_i), \tag{9a}$$

$$\mathbf{u}_i^{(t)} = \mathbf{b}_i^{(t)} + [\tilde{\mathcal{J}} \circ \tilde{\mathcal{G}} \circ \mathcal{S}_{\alpha_t}(\mathcal{G} \circ \mathcal{J}(\mathbf{b}^{(t)}))]_i, \quad 1 \le i \le N_c \tag{9b}$$

Our proposed reconstruction network thus is composed of a prescribed T phases, where the tth phase performs the update of (9). With a zero initial $\{\mathbf{u}_i^{(0)}\}$ and partial k-space data $\{\mathbf{f}_i\}$ as input, the network performs the update (9) for $1 \le t \le T$ and finally outputs $\mathbf{u}^{(T)}$. This network serves as a solver of (2) and uses $\mathbf{u}^{(T)}$ as an approximation of the true solution $\mathbf{u}(\mathbf{f}; \Theta)$. Hence, the constraint in (3) is replaced by this network for every input data $\mathbf{f}^{[j]}$.

3.3   Network Architectures and Training

We set $\mathcal{J}$ as a convolutional network with $N_l = 4$ layers and each linear convolution of kernel size 3×3. The first $N_l - 1$ layers have $N_f = 64$ filter kernels,

Table 1. Quantitative measurements for reconstruction of Coronal FSPD data.

Method	PSNR	SSIM	RMSE
GRAPPA [6]	24.9251 ± 0.9341	0.4827 ± 0.0344	0.2384 ± 0.0175
SPIRiT [10]	28.3525 ± 1.3314	0.6509 ± 0.0300	0.1614 ± 0.0203
VN [7]	30.2588 ± 1.1790	0.7141 ± 0.0483	0.1358 ± 0.0152
DeepcomplexMRI [20]	36.6268 ± 1.9662	0.9094 ± 0.0331	0.0653 ± 0.0085
pMRI-Net	**37.8475 ± 1.2086**	**0.9212 ± 0.0236**	**0.0568 ± 0.0069**

and $N_f = 1$ in the last layer. Each layer follows an activation ReLU except for the last layer. The operator $\mathcal{G}$ being set as the same way except that $N_f = 32$ and kernel size is 9×9. Operators $\tilde{\mathcal{J}}$ and $\tilde{\mathcal{G}}$ are designed in symmetric structures as $\mathcal{J}$ and $\mathcal{G}$ respectively. We treat a complex tensor as a real tensor of doubled size, and apply convolution separately. More details on network structure are provided in Supplementary Material.

The training data $(\mathbf{f}, \hat{\mathbf{u}})$ consists of J pairs $\{(\mathbf{f}_i^{[j]}, \hat{\mathbf{u}}_i^{[j]}) \mid 1 \leq i \leq N_c, \, 1 \leq j \leq J\}$. To increase network capacity, we allow varying operators of (9) in different phases. Hence $\Theta = \{\rho_t, \alpha_t, \mathcal{J}^{(t)}, \mathcal{G}^{(t)}, \tilde{\mathcal{G}}^{(t)}, \tilde{\mathcal{J}}^{(t)} \mid 1 \leq t \leq T\}$ are the parameters to be trained. Based on the analysis of loss functions [8,25,26], the optimal parameter Θ can be solved by minimizing the loss function: We set the discrepancy measure ℓ between the reconstruction $\mathbf{u}$ and the corresponding ground truth $\hat{\mathbf{u}}$ in (3) as follows,

$$\ell(\mathbf{u}, \hat{\mathbf{u}}) = \|\mathbf{s}(\mathbf{u}) - \mathbf{s}(\hat{\mathbf{u}})\|_2 + \gamma \||\mathcal{J}(\mathbf{u})| - \mathbf{s}(\hat{\mathbf{u}})\|_2 \tag{10}$$

where $\mathbf{s}(\mathbf{u}) = (\sum_{i=1}^{N_c} |\mathbf{u}_i|^2)^{1/2} \in \mathbb{R}^{mn}$ is the pointwise root of sum of squares across the N_c channels of $\mathbf{u}$, $|\cdot|$ is the pointwise modulus, and $\gamma > 0$ is a weight function. We also tried replacing the first by $(1/2) \cdot \|\mathbf{u} - \hat{\mathbf{u}}\|_2^2$, but it seems that the one given in (10) yields better results in our experiments. The second term of (10) can further improve accuracy of the magnitude of the reconstruction. The initial guess (also the input of the reconstruction network) of any given pMRI $\mathbf{f}^{[j]}$ is set to the zero-filled reconstruction $\mathbf{F}^{-1}\mathbf{f}^{[j]}$, and the multi-channel image $\mathbf{u}^{(T)}(\mathbf{f}^{[j]}; \Theta)$ is the output of the network (9) after T phases. In addition, $\mathcal{J}(\mathbf{u}^{(T)}(\mathbf{f}^{[j]}; \Theta))$ is the final single body image reconstructed as a by-product (complex-valued).

4   Experimental Results

Data. Two sequences of data named Coronal proton-density (PD) and Coronal fat-saturated proton-density (FSPD) along with the regular Cartesian sampling mask with 31.56% sampling ratio were obtained from https://github. com/VLOGroup/mri-variationalnetwork in our experiment. Each of the two sequences data were scanned from 20 patients. The training data consists of

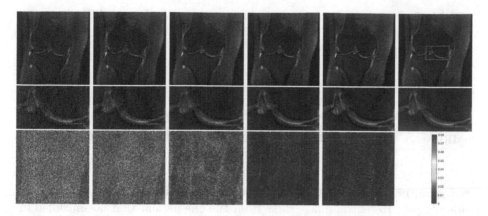

Fig. 1. Results on the Coronal FSPD knee image with regular Cartesian sampling (31.56% rate). From left to right columns: GRAPPA(25.6656/0.4671/0.2494), SPIRiT(29.5550/0.6574/0.1594), VN (31.5546/0.7387/0.1333), deepcomplexMRI (38.6842/0.9360/0.0587), pMRI-Net (38.8749/0.9375/0.0574), and ground truth (PSNR/SSIM/RMSE). From top to bottom rows: image, zoom-in views, and pointwise absolute error to ground truth.

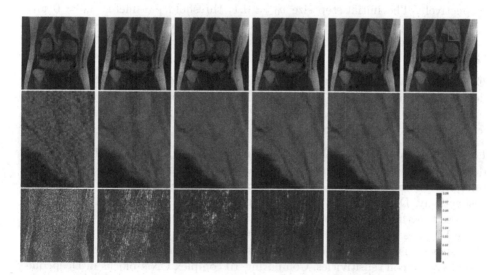

Fig. 2. Results on the Coronal PD knee image with regular Cartesian sampling (31.56% rate). From left to right columns: GRAPPA(29.9155/0.7360/0.1032), SPIRiT(33.2350/0.8461/0.0704), VN (38.3192/0.9464/0.0393), DeepcomplexMRI (41.2098/0.9713/0.0281), pMRI-Net (42.9330/0.9798/0.0231), and ground truth (PSNR/SSIM/RMSE). From top to bottom rows: image, zoom-in views, and pointwise absolute error to ground truth.

Table 2. Quantitative measurements for reconstruction of Coronal PD data.

Method	PSNR	SSIM	RMSE
GRAPPA [6]	30.4154 ± 0.5924	0.7489 ± 0.0207	0.0984 ± 0.0030
SPIRiT [10]	32.0011 ± 0.7920	0.7979 ± 0.0306	0.0824 ± 0.0082
VN [7]	37.8265 ± 0.4000	0.9281 ± 0.0114	0.0422 ± 0.0036
DeepcomplexMRI [20]	41.5756 ± 0.6271	0.9679 ± 0.0031	0.0274 ± 0.0018
pMRI-Net	**42.4333 ± 0.8785**	**0.9793 ± 0.0023**	**0.0249 ± 0.0024**

526 central image slices with matrix size 320×320 from 19 patients, and we randomly pick 15 central image slices from the one patient that not included in training data as the testing data. We normalized training data by the maximum of the absolute valued zero-filled reconstruction.

Implementation. The proposed network was implemented with $T = 5$ phases. We use Xavier initialization [5] to initialize network parameters and Adam optimizer for training. Experiments apply mini-batches of 2 and 3000 epochs with learning rate 0.0001 and 0.0005 for training Coronal FSPD data and PD data respectively. The initial step size $\rho_0 = 0.1$, threshold parameter $\alpha_0 = 0$ and $\gamma = 10^5$ in the loss function. All the experiments were implemented in Tensor-Flow on a workstation with Intel Core i9-7900 CPU and Nvidia GTX-1080Ti GPU.

Evaluation. We evaluate traditional methods GRAPPA [6], SPIRiT [10], and deep learning methods VN [7], DeepcomplexMRI [20] over the 15 testing Coronal FSPD and PD knee images in terms of PSNR, SSIM [21] and RMSE (RMSE of $\hat{\mathbf{x}}$ to true $\mathbf{x}^*$ is defined by $\|\hat{\mathbf{x}} - \mathbf{x}^*\|/\|\mathbf{x}^*\|$).

Experimental Results. The average numerical performance with standard deviations are summarized in Table 1 and 2. The comparison on reconstructed images are shown in Fig. 1 and Fig. 2 for Coronal FSPD and PD testing data respectively. Despite of the lack of coil sensitivities in training and testing, the proposed method still outperforms VN in reconstruction accuracy significantly while VN uses precomputed coil sensitivity maps from ESPIRiT [17], which further shows that the proposed method can achieve improved accuracy without knowledge of coil sensitivities. Comparing 10 complex CNN blocks in DeepcomplexMRI with 5 phases in pMRI-Net, the latter requires fewer network parameters and less training time but improves reconstruction quality.

In the experiment of GRAPPA and SPIRiT, we use calibration kernel size 5×5 with Tikhonov regularization in the calibration setted as 0.01. We implement SPIRiT with 30 iterations and set Tikhonov regularization in the reconstruction as 10^{-3}. Default parameter settings for experiments of VN and DeepcomplexMRI were applied. The final recovered image from VN is a full FOV single channel image, and DeepcomplexMRI produces a multi-coil image, which are combined into single channel image using adaptive multi-coil combination method [18]. pMRI-Net reconstructs both single channel image $\mathcal{J}(\mathbf{u}^{(T)}(\mathbf{f}; \Theta))$ and multi-channel image $\{\mathbf{u}_i^{(T)}(\mathbf{f}_i; \Theta)\}$.

5 Conclusion

We exploit a learning based multi-coil MRI reconstruction without explicit knowledge of coil sensitivity maps and the network is modeled in CS framework with proximal gradient scheme. The proposed network is designed to combine features of channel-wise images, and then extract sparse features from the coil combined image. Our experiments showed better performance of proposed "combine-then-regularize" approach.

References

1. Aggarwal, H.K., Mani, M.P., Jacob, M.: MoDL: model-based deep learning architecture for inverse problems. IEEE Trans. Med. Imaging **38**(2), 394–405 (2019)
2. Cheng, J., Wang, H., Ying, L., Liang, D.: Model learning: primal dual networks for fast MR imaging. In: Shen, D., et al. (eds.) MICCAI 2019. LNCS, vol. 11766, pp. 21–29. Springer, Cham (2019). https://doi.org/10.1007/978-3-030-32248-9_3
3. Deshmane, A., Gulani, V., Griswold, M.A., Seiberlich, N.: Parallel MR imaging. J. Magn. Reson. Imaging **36**(1), 55–72 (2012)
4. Duan, J., et al.: VS-Net: variable splitting network for accelerated parallel MRI reconstruction. In: Shen, D., et al. (eds.) MICCAI 2019. LNCS, vol. 11767, pp. 713–722. Springer, Cham (2019). https://doi.org/10.1007/978-3-030-32251-9_78
5. Glorot, X., Bengio, Y.: Understanding the difficulty of training deep feedforward neural networks. In: Teh, Y.W., Titterington, M. (eds.) Proceedings of the Thirteenth International Conference on Artificial Intelligence and Statistics. Proceedings of Machine Learning Research, vol. 9, pp. 249–256. PMLR, Chia Laguna Resort, Sardinia, Italy, 13–15 May 2010
6. Griswold, M.A., et al.: Generalized autocalibrating partially parallel acquisitions (GRAPPA). Magn. Reson. Med. Official J. Int. Soc. Magn. Reson. Med. **47**(6), 1202–1210 (2002)
7. Hammernik, K., et al.: Learning a variational network for reconstruction of accelerated MRI data. Magn. Reson. Med. **79**(6), 3055–3071 (2018)
8. Hammernik, K., Knoll, F., Sodickson, D., Pock, T.: L2 or not L2: impact of loss function design for deep learning MRI reconstruction. In: Proceedings of the International Society of Magnetic Resonance in Medicine (ISMRM) (2017)
9. Kwon, K., Kim, D., Park, H.: A parallel MR imaging method using multilayer perceptron. Med. Phys. **44**(12), 6209–6224 (2017)
10. Lustig, M., Pauly, J.M.: SPIRiT: iterative self-consistent parallel imaging reconstruction from arbitrary k-space. Magn. Reson. Med. **64**(2), 457–471 (2010). https://doi.org/10.1002/mrm.22428. https://onlinelibrary.wiley.com/doi/abs/10.1002/mrm.22428
11. Mardani, M., et al.: Deep generative adversarial neural networks for compressive sensing MRI. IEEE Trans. Med. Imaging **38**(1), 167–179 (2019)
12. Meng, N., Yang, Y., Xu, Z., Sun, J.: A prior learning network for joint image and sensitivity estimation in parallel MR imaging. In: Shen, D., et al. (eds.) MICCAI 2019. LNCS, vol. 11767, pp. 732–740. Springer, Cham (2019). https://doi.org/10.1007/978-3-030-32251-9_80

13. Pruessmann, K.P., Weiger, M., Scheidegger, M.B., Boesiger, P.: Sense: sensitivity encoding for fast MRI. Magn. Reson. Med. Official J. Int. Soc. Magn. Reson. Med. **42**(5), 952–962 (1999)
14. Quan, T.M., Nguyen-Duc, T., Jeong, W.K.: Compressed sensing MRI reconstruction using a generative adversarial network with a cyclic loss. IEEE Trans. Med. Imaging **37**, 1488–1497 (2018)
15. Schlemper, J., Caballero, J., Hajnal, J.V., Price, A.N., Rueckert, D.: A deep cascade of convolutional neural networks for dynamic MR image reconstruction. IEEE Trans. Med. Imaging **37**(2), 491–503 (2018)
16. Sudarshan, V.P., Gupta, K., Egan, G., Chen, Z., Awate, S.P.: Joint reconstruction of PET + parallel-MRI in a Bayesian coupled-dictionary MRF framework. In: Shen, D., et al. (eds.) MICCAI 2019. LNCS, vol. 11766, pp. 39–47. Springer, Cham (2019). https://doi.org/10.1007/978-3-030-32248-9_5
17. Uecker, M., et al.: ESPIRiT—an eigenvalue approach to autocalibrating parallel MRI: where Sense meets Grappa. Magn. Reson. Med. **71**(3), 990–1001 (2014)
18. Walsh, D., Gmitro, A., Marcellin, M.: Adaptive reconstruction of phased array MR imagery. Magn. Reson. Med. **43**(5), 682–690 (2000). Cited By 363
19. Wang, S., e al.: Accelerating magnetic resonance imaging via deep learning. In: 2016 IEEE 13th International Symposium on Biomedical Imaging (ISBI), pp. 514–517, April 2016
20. Wang, S., et al.: DeepcomplexMRI: exploiting deep residual network for fast parallel MR imaging with complex convolution. J. Magn. Reson. Imaging **68**, 136–147 (2020)
21. Wang, Z., Bovik, A.C., Sheikh, H.R., Simoncelli, E.P.: Image quality assessment: from error visibility to structural similarity. IEEE Trans. Image Process. **13**(4), 600–612 (2004)
22. Yang, Y., Sun, J., Li, H., Xu, Z.: ADMM-CSNet: a deep learning approach for image compressive sensing. IEEE Trans. Pattern Anal. Mach. Intell. **42**(3), 521–538 (2020)
23. Yang, Y., Sun, J., Li, H., Xu, Z.: Deep ADMM-Net for compressive sensing MRI. In: Lee, D.D., Sugiyama, M., Luxburg, U.V., Guyon, I., Garnett, R. (eds.) Advances in Neural Information Processing Systems, vol. 29, pp. 10–18. Curran Associates, Inc. (2016)
24. Zhang, J., Ghanem, B.: ISTA-Net: interpretable optimization-inspired deep network for image compressive sensing. In: Proceedings of the IEEE Conference on Computer Vision and Pattern Recognition, pp. 1828–1837 (2018)
25. Zhao, H., Gallo, O., Frosio, I., Kautz, J.: Loss functions for image restoration with neural networks. IEEE Trans. Comput. Imaging **3**(1), 47–57 (2017)
26. Zhou, Z., et al.: Parallelimaging and convolutional neural network combined fast MR image reconstruction: applications in low-latency accelerated real-time imaging. Med. Phys. **46**(8), 3399–3413 (2019). https://doi.org/10.1002/mp.13628. https://aapm.onlinelibrary.wiley.com/doi/abs/10.1002/mp.13628

Neural Network-Based Reconstruction in Compressed Sensing MRI Without Fully-Sampled Training Data

Alan Q. Wang[1(✉)], Adrian V. Dalca[2,3], and Mert R. Sabuncu[1,4]

[1] School of Electrical and Computer Engineering, Cornell University, Ithaca, USA
aw847@cornell.edu
[2] Computer Science and Artificial Intelligence Lab at the Massachusetts Institute of Technology, Cambridge, USA
[3] A.A. Martinos Center for Biomedical Imaging at the Massachusetts General Hospital, Boston, USA
[4] Meinig School of Biomedical Engineering, Cornell University, Ithaca, USA

Abstract. Compressed Sensing MRI (CS-MRI) has shown promise in reconstructing under-sampled MR images, offering the potential to reduce scan times. Classical techniques minimize a regularized least-squares cost function using an expensive iterative optimization procedure. Recently, deep learning models have been developed that model the iterative nature of classical techniques by unrolling iterations in a neural network. While exhibiting superior performance, these methods require large quantities of ground-truth images and have shown to be non-robust to unseen data. In this paper, we explore a novel strategy to train an unrolled reconstruction network in an unsupervised fashion by adopting a loss function widely-used in classical optimization schemes. We demonstrate that this strategy achieves lower loss and is computationally cheap compared to classical optimization solvers while also exhibiting superior robustness compared to supervised models. Code is available at https://github.com/alanqrwang/HQSNet.

Keywords: Compressed sensing MRI · Unsupervised reconstruction · Model robustness

1 Introduction

Magnetic resonance (MR) imaging can be accelerated via under-sampling k-space – a technique known as Compressed Sensing MRI (CS-MRI) [25]. This yields a well-studied ill-posed inverse problem. Classically, this problem is reduced to regularized regression, which is solved via an iterative optimization scheme, e.g., [5,7,8,11,14,38], on each collected measurement set. The limitations of this instance-based optimization approach are well-known; solutions are heavily influenced by the choice of regularization function and can lack in high-frequency detail. Furthermore, they are often time-consuming to compute.

© Springer Nature Switzerland AG 2020
F. Deeba et al. (Eds.): MLMIR 2020, LNCS 12450, pp. 27–37, 2020.
https://doi.org/10.1007/978-3-030-61598-7_3

There has been a recent surge in deep learning methods for CS-MRI, which promise superior performance and computational efficiency. To date, these methods have largely relied on fully-sampled data, which are under-sampled retrospectively to train the neural network model. The primary focus of this body of research has been the design of the neural network architecture. So-called "unrolled architectures" [1,20,24,26,32,34,37] which inject the MR-specific forward model into the network architecture have been shown to outperform more general-purpose, black-box [23,24,35] models and aforementioned classical methods.

While these methods exhibit state-of-the-art reconstruction performance, a major limitation of the supervised formulation is the necessity for a dataset of fully-sampled ground-truth images, which can be hard to obtain in the clinical setting. In addition, these models are known to exhibit poor robustness at test-time when subjected to noisy perturbations and adversarial attacks [2].

In this work, we present a novel approach for performing MR reconstruction that combines the robustness of classical techniques with the performance of unrolled architectures. Specifically, we implement an unrolled neural network architecture that is trained to minimize a classical loss function, which does not rely on fully-sampled data. This is an "amortized optimization" of the classical loss, and we refer to our model as "unsupervised". We show that our unsupervised model can be more robust than its supervised counterpart under noisy scenarios. Additionally, we demonstrate that not only can we replace an expensive iterative optimization procedure with a simple forward pass of a neural network, but also that this method can outperform classical methods even when trained to minimize the same loss.

2   Background

In the CS-MRI formulation, fully-sampled MR images are assumed to be transformed into under-sampled k-space measurements by the forward model:

$$y = \mathcal{F}_\Omega x, \tag{1}$$

where $x \in \mathbb{C}^N$ is the unobserved fully-sampled image, $y \in \mathbb{C}^M$ is the under-sampled k-space measurement vector[1], $M < N$, and $\mathcal{F}_\Omega$ denotes the under-sampled Fourier operator with Ω indicating the index set over which the k-space measurements are sampled. For each instance y, classical methods solve the ill-posed inverse problem via an optimization of the form:

$$\arg\min_x \|\mathcal{F}_\Omega x - y\|_2^2 + \mathcal{R}(x), \tag{2}$$

where $\mathcal{R}(x)$ denotes a regularization loss term. The regularization term is often carefully engineered to restrict the solutions to the space of desirable images. Common choices include sparsity-inducing norms of wavelet coefficients [16],

---

[1] In this paper, we assume a single coil acquisition.

total variation [21, 31], and their combinations [25, 30]. The first term of Eq. (2), called the data consistency term, quantifies the agreement between the measurement vector y and reconstruction x.

Half-quadratic splitting (HQS) [18, 28] solves Eq. (2) by decoupling the minimization of the two competing terms using an auxiliary variable z and an alternating minimization strategy over iterations $k \in \mathbb{N}$:

$$z_k = \arg\min_z \mathcal{R}(z) + \lambda \|z - x_k\|_2^2, \tag{3a}$$

$$x_{k+1} = \arg\min_x \|\mathcal{F}_\Omega x - y\|_2^2 + \lambda \|z_k - x\|_2^2, \tag{3b}$$

where $\lambda \geq 0$ is a hyper-parameter. Equation (3b) has closed-form solution in k-space at the sampling location m given by:

$$\hat{x}_{k+1}[m] = \begin{cases} \frac{y[m] + \lambda \hat{z}_k[m]}{1 + \lambda}, & \text{if } m \in \Omega \\ \hat{z}_k[m], & \text{else} \end{cases} \tag{4}$$

for all k, where $\hat{x}_{k+1}$ and $\hat{z}_k$ denote x_{k+1} and z_k in Fourier domain, respectively. The z-minimization of Eq. (3a) is the proximal operator for the regularizer, which may be solved using (sub-)gradient descent for differentiable $\mathcal{R}$. In this paper, we view the proximal operator as a function, i.e. $z_k = g(x_{k+1})$, where g is some neural network. HQS and its data-driven variants underlie algorithms in CS-MRI [1, 32], image super-resolution [10], and image restoration [15].

Supervised deep learning offers an alternative approach. Given a dataset of fully-sampled images and (often retrospectively-created) under-sampled measurements $\mathcal{D} = \{(x_i, y_i)\}_{i=1}^N$, these methods learn the optimal parameters θ of a parameterized mapping $G_\theta : y_i \to x_i$ by minimizing:

$$\arg\min_\theta \frac{1}{N} \sum_{i=1}^N \mathcal{L}_{sup}(G_\theta(y_i), x_i), \tag{5}$$

where $\mathcal{L}_{sup}$ is a loss function that quantifies the quality of reconstructions based on the fully-sampled x. This formulation obviates the need for the design of a regularization loss function. The parameterized mapping is often a neural network model [3, 23, 24, 35]. Recently, unrolled architectures that exploit knowledge of the forward model [1, 26, 32] have proven to be effective. These architectures implement layers that iteratively minimize the data consistency loss and remove aliasing artifacts by learning from fully-sampled data. $K \in \mathbb{N}$ such blocks are concatenated and trained end-to-end to minimize the supervised loss of Eq. (5).

3 Proposed Method

To remove the need for fully-sampled data, leverage the robustness of the classical optimization method, and incorporate the performance of deep learning models, we propose to use an unsupervised strategy with an unrolled reconstruction architecture, which we call HQS-Net. Let a parameterized mapping G_θ denote a neural network that maps under-sampled measurements to reconstructed images. We train this network to minimize over θ:

$$\mathcal{L}(y_i; \theta) = \frac{1}{N} \sum_{i=1}^{N} \left[\|\mathcal{F}_\Omega G_\theta(y_i) - y_i\|_2^2 + \mathcal{R}\left(G_\theta(y_i)\right) \right]. \tag{6}$$

This model can be viewed as an amortization of the instance-specific optimization of Eq. (2), via a neural network G_θ [4,12,19,27,33].

Amortized optimization provides several advantages over classical solutions. First, at test-time, it replaces an expensive iterative optimization procedure with a simple forward pass of a neural network. Second, since the function G_θ is tasked with estimating the reconstruction for any viable input measurement vector y and not just a single instance, amortized optimization has been shown to act as a natural regularizer for the optimization problem [4,33].

3.1   Model Architecture

Similar to instance-based iterative procedures like HQS and supervised unrolled architectures such as [1,32], HQS-Net decouples the minimization of the data consistency term and regularization term in Eq. (6). Specifically, in each iteration block, the network explicitly enforces data consistency preceded by a convolutional block g_{θ_k} that learns an iteration-specific regularization. Thus, we obtain an alternating minimization analogous to Eq. (3):

$$z_k = g_{\theta_k}\left(x_k\right), \tag{7a}$$

$$\hat{x}_{k+1}[m] = \begin{cases} \frac{y[m]+\lambda \hat{z}_k[m]}{1+\lambda}, & \text{if } m \in \Omega \\ \hat{z}_k[m], & \text{else} \end{cases} \tag{7b}$$

where $x_1 = \mathcal{F}_\Omega^H y$ (i.e. the zero-filled reconstruction). Equation (7b) is implemented as a data-consistency layer (DC) within the network[2]. The unrolled network concatenates $K \in \mathbb{N}$ of these g_{θ_k} and DC blocks, as shown in Fig. 1.

4   Experiments

In our experiments, we used three different MRI datasets: T1-weighted axial brain, T2-weighted axial brain, and PD-weighted coronal knee scans. We applied retrospective down-sampling with 4-fold and 8-fold acceleration sub-sampling masks generated using a Poisson-disk variable-density sampling strategy [9,17,

---

[2] For forward models that do not permit an analytical solution of Eq. (7b) (e.g. multi-coil MRI), one can replace the data-consistency layer with an iterative optimization scheme (e.g. conjugate gradient as in [1]). In addition, the iteration-specific weights θ_k in Eq. (7a) can be replaced by a shared set of weights θ, which enforces the model to learn a global regularization prior for all iterations.

Fig. 1. Proposed architecture. CNN and DC layers are unrolled K times as a deep network, and the final output $G_\theta = x_K$ is encouraged to minimize $\mathcal{L}$ defined in Eq. (6). $\mathcal{L}$ does not see fully-sampled data x.

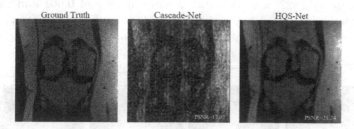

Fig. 2. Example of typical reconstructions of PD-weighted knee for 8-fold acceleration under additive white Gaussian spatial noise, where noise standard deviation $\sigma = 0.1$.

25]. We use 2nd-order and 3rd-order polynomial densities for the 4-fold and 8-fold masks, respectively. All training and testing experiments in this paper were performed on a machine equipped with an Intel Xeon Gold 6126 processor and an NVIDIA Titan Xp GPU.

Data. T1-weighted brain scans were obtained from [13], T2-weighted brain scans were obtained from the IXI dataset[3], and PD-weighted knee scans were obtained from the fastMRI NYU dataset [39]. All images were intensity-normalized to the range $[0, 1]$ and cropped and re-sampled to a pixel grid of size 256×256. Dataset sizes consisted of 2000, 500, and 1000 slices for training, validation, and testing, respectively.

Comparison Models. We compared the proposed HQS-Net against HQS and Cascade-Net [32], a supervised upper-bound baseline model. All models were implemented in Pytorch.

HQS minimizes the instance-based loss in Eq. (2) using the alternating algorithm in Eq. (3), where the z-minimization is performed using gradient descent. All iterative procedures were run until convergence within a specified tolerance. In choosing $\mathcal{R}(x)$, we followed the literature [25, 30] and let

$$\mathcal{R}(x) = \alpha TV(x) + \beta \|Wx\|_1, \tag{8}$$

where TV denotes total variation, W denotes the discrete wavelet transform operator, $\alpha, \beta > 0$ are weighting coefficients, and $\|\cdot\|_1$ denotes the ℓ_1 norm.

---

[3] https://brain-development.org/ixi-dataset.

Cascade-Net is trained to minimize Eq. (5) using an ℓ_2 loss and with an identical model architecture as HQS-Net. Thus, Cascade-Net requires access to fully-sampled training data, whereas HQS and HQS-Net do not.

For g_{θ_k}, a 5-layer model was used with channel size 64 at each layer. Each layer consists of convolution followed by a ReLU activation function. We used a residual learning strategy which adds the zero-filled input to the output of the CNN. The overall architecture G_θ is unrolled such that $K = 25$. For training, Adam optimization [22] was used with a learning rate of 0.001 and batch size of 8. In experiments, we set $\lambda = 1.8$, $\alpha = 0.005$, and $\beta = 0.002$, which were optimized using a Bayesian Optimization hyper-parameter tuning package [6].

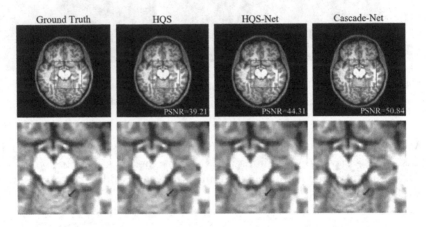

Fig. 3. Example reconstructions of a T1-weighted axial brain slice for 4-fold acceleration. Although classical and unsupervised methods minimize the same loss, the unsupervised model is able to retain more high-frequency detail.

Evaluation Metrics. Reconstructions were evaluated against ground-truth images on peak signal-to-noise ratio (PSNR), structural similarity index (SSIM) [36], and high-frequency error norm (HFEN) [29]. The *relative* value (e.g., relative PSNR) for a given reconstruction was computed by subtracting the corresponding metric value for the zero-filled reconstruction.

4.1 Results

Runtime and Loss Analysis. Table 1 shows the average runtime and average loss value (defined in Eq. (2) and (8)) achieved by both HQS and HQS-Net on the test set. Since inference for HQS-Net equates to a forward pass through the trained network, HQS-Net is several orders of magnitude faster while achieving superior loss compared to HQS.

Table 1. Inference runtime and loss. Lower is better. Mean ± standard deviation across test cases. 4-fold acceleration.

Dataset	Method	Inference time (sec)	Loss
T1 brain	HQS	483 ± 111	17.34 ± 2.94
	HQS-Net	0.241 ± 0.013	17.20 ± 2.99
T2 brain	HQS	380 ± 72	19.93 ± 4.20
	HQS-Net	0.246 ± 0.023	19.46 ± 4.13
PD knee	HQS	366 ± 129	25.28 ± 10.93
	HQS-Net	0.251 ± 0.013	24.61 ± 10.58

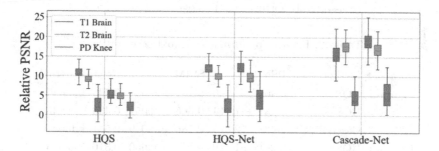

Fig. 4. Reconstruction performance for comparison models of all datasets evaluated on relative PSNR for 4-fold (left) and 8-fold (right) acceleration rates.

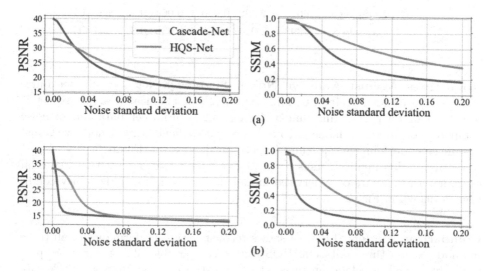

Fig. 5. Average reconstruction performance vs. noise standard deviation for 8-fold acceleration, evaluated on PSNR and SSIM. (a) shows performance under additive noise in image domain and (b) shows performance under additive noise in k-space.

Table 2. Relative performance. Higher is better. Mean ± standard deviation across test cases. 4-fold acceleration.

Dataset	Method	PSNR	SSIM	Negative HFEN
T1 brain	HQS	10.88 ± 1.256	0.401 ± 0.052	0.154 ± 0.016
	HQS-Net	12.10 ± 1.431	0.409 ± 0.052	0.157 ± 0.016
	Cascade-Net	15.72 ± 2.489	0.416 ± 0.051	0.180 ± 0.016
T2 brain	HQS	9.266 ± 0.961	0.369 ± 0.057	0.172 ± 0.014
	HQS-Net	10.00 ± 1.055	0.382 ± 0.057	0.177 ± 0.015
	Cascade-Net	17.64 ± 1.910	0.399 ± 0.056	0.209 ± 0.017
PD knee	HQS	2.387 ± 2.120	0.016 ± 0.021	0.113 ± 0.057
	HQS-Net	2.472 ± 2.117	0.018 ± 0.021	0.120 ± 0.054
	Cascade-Net	4.419 ± 2.081	0.086 ± 0.023	0.179 ± 0.048

Table 3. Relative performance. 8-fold acceleration

Dataset	Method	PSNR	SSIM	Negative HFEN
T1 brain	HQS	5.403 ± 1.500	0.161 ± 0.055	0.303 ± 0.045
	HQS-Net	12.31 ± 1.814	0.566 ± 0.052	0.436 ± 0.035
	Cascade-Net	15.38 ± 2.570	0.586 ± 0.049	0.531 ± 0.025
T2 brain	HQS	5.261 ± 1.520	0.211 ± 0.050	0.310 ± 0.055
	HQS-Net	9.880 ± 1.839	0.553 ± 0.052	0.457 ± 0.031
	Cascade-Net	16.92 ± 1.898	0.589 ± 0.044	0.576 ± 0.025
PD knee	HQS	2.109 ± 1.277	0.051 ± 0.040	0.199 ± 0.094
	HQS-Net	4.019 ± 2.782	0.085 ± 0.063	0.252 ± 0.101
	Cascade-Net	5.393 ± 3.043	0.156 ± 0.038	0.347 ± 0.103

Robustness Against Additive Noise. Since HQS-Net does not see fully-sampled data and is trained to minimize a robust classical loss, we expect it to exhibit better performance under circumstances of unseen data and/or noise compared to supervised models. To test this, we artificially inject additive Gaussian noise in both image space and k-space on the test set. Figure 5 shows a plot of reconstruction quality versus noise variance for 8-fold acceleration. A visual example of the failure of supervised models to perform well under noisy conditions is shown in Fig. 2.

Comparison Models. Figure 4 shows reconstruction performance of all three methods across three datasets. HQS-Net is comparable, if not superior (particularly at high acceleration rates), to the instance-based HQS solver despite the fact that they optimize the same loss function. This may be attributed to the network being trained across many samples, such that it is able to leverage commonalities in structure and detail across the entire training set. Figure 3 highlights the learning of high-frequency detail (Table 2 and 3).

5 Conclusion

We explored a novel unsupervised MR reconstruction method that performs an amortized optimization of the classical loss formulation for CS-MRI, thus eliminating the need for fully-sampled ground-truth data. We show that our method is more robust to noise as compared to supervised methods that have the same network architecture and is computationally cheaper than classical solvers that minimize the same loss. While our experiments focused on MRI, the method is broadly applicable to other imaging modalities and can be improved with more expressive networks and/or regularization functions.

Acknowledgements. This research was funded by NIH grants R01LM012719, R01AG053949; and, NSF CAREER 1748377, and NSF NeuroNex Grant1707312.

References

1. Aggarwal, H.K., Mani, M.P., Jacob, M.: MoDL: model-based deep learning architecture for inverse problems. IEEE Trans. Med. Imaging **38**(2), 394–405 (2019)
2. Antun, V., Renna, F., Poon, C., Adcock, B., Hansen, A.C.: On instabilities of deep learning in image reconstruction - does AI come at a cost? (2019)
3. Bahadir, C.D., Wang, A.Q., Dalca, A.V., Sabuncu, M.R.: Deep-learning-based optimization of the under-sampling pattern in MRI. IEEE Trans. Comput. Imaging **6**, 1139–1152 (2020)
4. Balakrishnan, G., Zhao, A., Sabuncu, M.R., Guttag, J., Dalca, A.V.: VoxelMorph: a learning framework for deformable medical image registration. IEEE Trans. Med. Imaging **38**(8), 1788–1800 (2019)
5. Beck, A., Teboulle, M.: A fast iterative shrinkage-thresholding algorithm for linear inverse problems. SIAM J. Imging Sci. **2**(1), 183–202 (2009)
6. Bergstra, J., Yamins, D., Cox, D.D.: Making a science of model search: hyperparameter optimization in hundreds of dimensions for vision architectures. In: Proceedings of the 30th International Conference on International Conference on Machine Learning, ICML 2013, vol. 28. pp. I-115–I-123. JMLR.org
7. Boyd, S., Parikh, N., Chu, E., Peleato, B., Eckstein, J.: Distributed optimization and statistical learning via the alternating direction method of multipliers. Found. Trends Mach. Learn. **3**(1), 1–122 (2011)
8. Chambolle, A., Pock, T.: A first-order primal-dual algorithm for convex problems with applications to imaging. J. Math. Imaging Vis. **40**, 120–145 (2011)
9. Chauffert, N., Ciuciu, P., Weiss, P.: Variable density compressed sensing in MRI. theoretical vs heuristic sampling strategies. In: 2013 IEEE 10th International Symposium on Biomedical Imaging, April 2013
10. Cheng, K., Du, J., Zhou, H., Zhao, D., Qin, H.: Image super-resolution based on half quadratic splitting. Infrared Phys. Technol. **105**, 103193 (2020)

11. Combettes, P.L., Pesquet, J.C.: Proximal splitting methods in signal processing. In: Bauschke, H., Burachik, R., Combettes, P., Elser, V., Luke, D., Wolkowicz, H. (eds.) Fixed-Point Algorithms for Inverse Problems in Science and Engineering, pp. 185–212. Springer, New York (2011). https://doi.org/10.1007/978-1-4419-9569-8_10

12. Cremer, C., Li, X., Duvenaud, D.: Inference suboptimality in variational autoencoders (2018)

13. Dalca, A.V., Guttag, J., Sabuncu, M.R.: Anatomical priors in convolutional networks for unsupervised biomedical segmentation. In: The IEEE Conference on Computer Vision and Pattern Recognition (CVPR) (2018)

14. Daubechies, I., Defrise, M., Mol, C.D.: An iterative thresholding algorithm for linear inverse problems with a sparsity constraint. Commun. Pure Appl. Math. **57**, 1413–1457 (2004)

15. Dong, W., Wang, P., Yin, W., Shi, G., Wu, F., Lu, X.: Denoising prior driven deep neural network for image restoration. IEEE Trans. Pattern Anal. Mach. Intell. **41**(10), 2305–2318 (2018)

16. Figueiredo, M.A.T., Nowak, R.D.: An EM algorithm for wavelet-based image restoration. IEEE Trans. Image Process. **12**(8), 906–916 (2003)

17. Geethanath, S., et al.: Compressed sensing MRI: a review. Crit. Rev. Biomed. Eng. **41**(3), 183–204 (2013)

18. Geman, D., Yang, C.: Nonlinear image recovery with half-quadratic regularization. IEEE Trans. Image Process. **4**(7), 932–946 (1995)

19. Gershman, S.J., Goodman, N.D.: Amortized inference in probabilistic reasoning. In: CogSci (2014)

20. Hammernik, K., et al.: Learning a variational network for reconstruction of accelerated MRI data. Magn. Reson. Med. **79**(6), 3055–3071 (2017)

21. Hu, Y., Jacob, M.: Higher degree total variation (HDTV) regularization for image recovery. IEEE Trans. Image Process. **21**(5), 2559–2571 (2012)

22. Kingma, D.P., Ba, J.: Adam: a method for stochastic optimization (2014)

23. Lee, D., Yoo, J., Ye, J.C.: Deep residual learning for compressed sensing MRI. In: 2017 IEEE 14th International Symposium on Biomedical Imaging (ISBI 2017), pp. 15–18, April 2017

24. Liang, D., Cheng, J., Ke, Z., Ying, L.: Deep MRI reconstruction: unrolled optimization algorithms meet neural networks (2019)

25. Lustig, M., Donoho, D., Pauly, J.M.: Sparse MRI: the application of compressed sensing for rapid MR imaging. Magn. Reson. Med. **58**(6), 1182–1195 (2007)

26. Mardani, M., et al.: Neural proximal gradient descent for compressive imaging (2018)

27. Marino, J., Yue, Y., Mandt, S.: Iterative amortized inference (2018)

28. Nikolova, M., Ng, M.K.: Analysis of half-quadratic minimization methods for signal and image recovery. SIAM J. Sci. Comput. **27**(3), 937–966 (2005)

29. Ravishankar, S., Bresler, Y.: MR image reconstruction from highly undersampled k-space data by dictionary learning. IEEE Trans. Med. Imaging **30**(5), 1028–1041 (2011)

30. Ravishankar, S., Ye, J.C., Fessler, J.A.: Image reconstruction: from sparsity to data-adaptive methods and machine learning. Proc. IEEE **108**(1), 86–109 (2020)

31. Rudin, L.I., Osher, S., Fatemi, E.: Nonlinear total variation based noise removal algorithms. Physica D **60**(1–4), 259–268 (1992)

32. Schlemper, J., Caballero, J., Hajnal, J.V., Price, A., Rueckert, D.: A deep cascade of convolutional neural networks for MR image reconstruction. In: Niethammer, M., et al. (eds.) IPMI 2017. LNCS, vol. 10265, pp. 647–658. Springer, Cham (2017). https://doi.org/10.1007/978-3-319-59050-9_51
33. Shu, R., Bui, H.H., Zhao, S., Kochenderfer, M.J., Ermon, S.: Amortized inference regularization (2018)
34. Tezcan, K.C., Baumgartner, C.F., Luechinger, R., Pruessmann, K.P., Konukoglu, E.: MR image reconstruction using deep density priors. IEEE Trans. Med. Imaging **38**(7), 1633–1642 (2019)
35. Wang, S., et al.: Accelerating magnetic resonance imaging via deep learning. In: 2016 IEEE 13th International Symposium on Biomedical Imaging (ISBI), pp. 514–517, April 2016
36. Wang, Z., Bovik, A.C., Sheikh, H.R., Simoncelli, E.P.: Image quality assessment: from error visibility to structural similarity. IEEE Trans. Imaging Process. **13**(4), 600–612 (2004)
37. Yang, Y., Sun, J., Li, H., Xu, Z.: ADMM-Net: a deep learning approach for compressive sensing MRI (2017)
38. Ye, N., Roosta-Khorasani, F., Cui, T.: Optimization methods for inverse problems. 2017 MATRIX Annals. MBS, vol. 2, pp. 121–140. Springer, Cham (2019). https://doi.org/10.1007/978-3-030-04161-8_9
39. Zbontar, J., et al.: fastMRI: an open dataset and benchmarks for accelerated MRI (2018)

Deep Recurrent Partial Fourier Reconstruction in Diffusion MRI

Fasil Gadjimuradov[1,2]($\boxtimes$), Thomas Benkert[2], Marcel Dominik Nickel[2], and Andreas Maier[1]

[1] Pattern Recognition Lab, Department of Computer Science,
Friedrich-Alexander-Universität Erlangen-Nürnberg, Erlangen, Germany
fasil.gadjimuradov@fau.de
[2] MR Application Development, Siemens Healthcare GmbH, Erlangen, Germany

Abstract. Partial Fourier (PF) acquisition schemes are often employed to increase the inherently low SNR in diffusion-weighted (DW) images. The resulting ill-posed reconstruction problem can be tackled by an iterative Projection Onto Convex Sets (POCS). By relaxing the data constraint and replacing the heuristically chosen regularization by learned convolutional filters, we arrive at an unrolled recurrent network architecture which circumvents weaknesses of the conventional POCS. Further, knowledge on the pixel-wise noise level of MR images is incorporated into data consistency operations within the reconstruction network. We are able to demonstrate on DW images of the pelvis that the proposed model quantitatively and qualitatively outperforms conventional methods as well as a U-Net representing a direct image-to-image mapping.

Keywords: Deep learning · Partial Fourier · Diffusion MRI

1 Introduction

Diffusion-weighted imaging (DWI) has emerged as a valuable instrument for localizing and characterizing abdominopelvic lesions in clinical MRI [14,20]. Because the encoding process of DWI involves the application of strong magnetic field gradients which spoil parts of the magnetization, image quality suffers from inherently low signal-to-noise ratio (SNR). One way to alleviate this problem from acquisition side is to employ Partial Fourier (PF) sampling that asymmetrically covers the k-space by omitting a contiguous part on one side. Since DW images are most commonly acquired by means of single-shot echo-planar imaging (EPI), PF sampling along the phase encoding direction allows to reduce the effective echo time (TE) and consequently increase SNR. However, the zero-filled reconstruction will suffer from blurring along the phase encoding direction. An effective and robust PF reconstruction method for DW images offers several

Electronic supplementary material The online version of this chapter (https://doi.org/10.1007/978-3-030-61598-7_4) contains supplementary material, which is available to authorized users.

benefits to clinicians as assessment and delineation of potential lesions could be facilitated through improved SNR. Alternatively, the number of excitations (NEX) per slice – and hence, scan time – could be reduced while achieving similar SNR as without PF sampling.

Conventional PF methods, such as Homodyne reconstruction [16] and Projection Onto Convex Sets (POCS) [8], rely on the assumption of a smooth image phase which is estimated from the symmetrically sampled part around the k-space center. However, since DW images are known to be subject to rapid phase variations, the likelihood of the smoothness assumption to be violated increases with higher asymmetry in k-space. Based on that, both Homodyne reconstruction and POCS may provide inaccurate reconstruction results for strong PF factors. Furthermore, both methods are sensitive with respect to noise in the measured data.

More recently, a convolutional neural network (CNN) trained in a supervised manner was applied to PF-MRI in [15]. Despite being trained on non-MR images from the ImageNet database [3], which have been retrospectively sub-sampled and equipped with a randomly simulated phase, a U-Net [18] was shown to successfully deblur and denoise zero-filled DW images of the brain. However, one major drawback of networks representing pure image-to-image mappings is the lack of guarantee of consistency with measured data which is crucial in the context of medical imaging.

Following previous contributions which emphasized the aspect of rendering deep networks more interpretable by using model-based architectures [1] and incorporating known operators [12], we propose to unroll a relaxed version of the iterative POCS algorithm into a network architecture which alternates between recurrent convolutional units, enforcing prior knowledge, and data consistency operations. Concerning the latter, knowledge on the pixel-wise noise level of images is incorporated. Further, the repetitive nature of DWI is exploited by processing multiple acquisitions of the same slice simultaneously using three-dimensional (3-D) convolutions. The proposed method is end-to-end trainable and does not require the manual selection of hyper-parameters during inference.

2   Methods

2.1   POCS and Proximal Splitting

In single-channel PF-MRI, the mapping from the image $x \in \mathbb{C}^N$ to its corresponding k-space measurements $y \in \mathbb{C}^M$ can be modeled as

$$y = Ax + n \tag{1}$$

where $A \in \mathbb{C}^{M \times N}$ is the linear forward operator, encoding both PF sub-sampling and discrete Fourier transform, and $n \in \mathbb{C}^M$ represents additive complex Gaussian noise. Due to the sub-sampling ($M < N$) and noise, recovering x from its measurements y is an ill-posed inverse problem.

The POCS algorithm tries to interpret the reconstruction as a convex feasibility problem requiring the solution to lie in the intersection of two nonempty convex sets Ω_1 and Ω_2 with the respective indicator functions defined as

$$\iota_{\Omega_1}(\boldsymbol{x}) = \begin{cases} 0, & \text{if } \angle \boldsymbol{x} = \phi \\ \infty, & \text{else} \end{cases} \quad \text{and} \quad \iota_{\Omega_2}(\boldsymbol{x}) = \begin{cases} 0, & \text{if } \boldsymbol{Ax} = \boldsymbol{y} \\ \infty, & \text{else} \end{cases}.$$

Here, Ω_1 encodes prior knowledge as it defines the set of all complex images that have a phase equal to the smooth phase estimate $\phi \in \mathbb{R}^N$, whereas Ω_2 comprises all images that are consistent with measured data $\boldsymbol{y}$. A solution to the convex feasibility problem can be found by applying the projections onto both sets, denoted as P_{Ω_1} and P_{Ω_2}, in an alternating fashion, resulting in the following iterative update rule with $k = 1, ..., K$:

$$\boldsymbol{z}_k = P_{\Omega_1}(\boldsymbol{x}_{k-1}) = \text{Re}(\boldsymbol{x}_{k-1} \cdot e^{-i\phi}) \cdot e^{i\phi} \tag{2a}$$

$$\boldsymbol{x}_k = P_{\Omega_2}(\boldsymbol{z}_k) = \boldsymbol{z}_k - \boldsymbol{A}^*(\boldsymbol{Az}_k - \boldsymbol{y}) \tag{2b}$$

where Re is the element-wise real part extractor and $\boldsymbol{A}^*$ is the adjoint operator of $\boldsymbol{A}$. The initial estimate is the zero-filled reconstruction computed as $\boldsymbol{x}_0 = \boldsymbol{A}^*\boldsymbol{y}$.

Instead of adhering to the constrained definition of the POCS which may be suboptimal in noisy settings, we adopt a more general approach by setting up the commonly used unconstrained optimization problem of the form

$$\arg\min_{\boldsymbol{x}} \mathcal{R}(\boldsymbol{x}) + \nu \mathcal{D}(\boldsymbol{Ax}, \boldsymbol{y}) \tag{3}$$

where $\mathcal{R}$ and $\mathcal{D}$ represent regularization and data terms, respectively, and $\nu > 0$ is a trade-off parameter weighting both terms against each other. Using a proximal splitting algorithm, (3) can be solved by iteratively applying the proximal operators on both $\mathcal{R}$ and $\mathcal{D}$ under an iteration-dependent step size $\mu_k \geq 0$:

$$\boldsymbol{z}_k = \text{prox}_{\mu_k \mathcal{R}}(\boldsymbol{x}_{k-1}) = \arg\min_{\boldsymbol{z}} \mathcal{R}(\boldsymbol{z}) + \frac{\mu_k}{2}||\boldsymbol{x}_{k-1} - \boldsymbol{z}||_2^2 \tag{4a}$$

$$\boldsymbol{x}_k = \text{prox}_{\mu_k \mathcal{D}}(\boldsymbol{z}_k) = \arg\min_{\boldsymbol{x}} \nu \mathcal{D}(\boldsymbol{Ax}, \boldsymbol{y}) + \frac{\mu_k}{2}||\boldsymbol{x} - \boldsymbol{z}_k||_2^2. \tag{4b}$$

In fact, as proximal operators constitute natural generalizations of projections, it can be easily seen that POCS is a special case of the proximal splitting algorithm with $\mathcal{R} = \iota_{\Omega_1}$ and $\mathcal{D} = \iota_{\Omega_2}$.

2.2   Noise Level-Dependent Data Consistency

As demonstrated in the supplementary material, a closed-form solution of (4b) for $\mathcal{D}(\boldsymbol{Ax}, \boldsymbol{y}) = \frac{1}{2}||\boldsymbol{Ax} - \boldsymbol{y}||_2^2$ can be found as

$$\boldsymbol{x}_k = \boldsymbol{z}_k - \frac{1}{1 + \lambda_k}\boldsymbol{A}^*(\boldsymbol{Az}_k - \boldsymbol{y}) \quad \text{with} \quad \lambda_k = \frac{\mu_k}{\nu}. \tag{5}$$

Parameter $\lambda_k \geq 0$ depends on the noise level as it controls the degree to which deviation from the original measurements is tolerated. For instance, (5) reduces to the hard projection in (2b) for the noise-less case with $\lambda_k = 0$. In order to control the effective strength of the data consistency operation, we propose to utilize a spatially varying, MR-specific noise map incorporating the magnetic bias field and the g-factor for parallel imaging [10]. Hence, we substitute λ_k with the product of a scaling factor w_k and the noise map represented by diagonal matrix $\boldsymbol{\Sigma} \in \mathbb{R}^{N \times N}$, turning (5) into

$$\boldsymbol{x}_k = \boldsymbol{z}_k - (\boldsymbol{I} + w_k \boldsymbol{\Sigma})^{-1} \boldsymbol{A}^* (\boldsymbol{A} \boldsymbol{z}_k - \boldsymbol{y}) \tag{6}$$

where $\boldsymbol{I} \in \mathbb{R}^{N \times N}$ denotes the identity matrix.

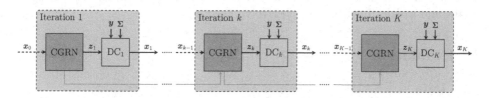

Fig. 1. Schematic overview of the unrolled network architecture. Within an iteration, the convolutional gated recurrent network (CGRN) enforcing prior knowledge on the current estimate is followed by a data consistency layer (DC). In order to adapt to iteration-specific requirements, the CGRN keeps track of an internal memory which is propagated across iterations (blue connections). (Color figure online)

2.3   Regularization via Convolutional Gated Recurrent Units

Being equipped with a way to compute the data consistency step, the task of finding a suitable regularizer remains. In the style of previous work, we choose to learn it from data by means of a CNN. One way to achieve this is to employ a "plug-and-play" approach in which a CNN is trained off-line and used to replace the regularization term in an arbitrary iterative algorithm, e.g. as shown in [7] and [13]. However, this has the disadvantage that the user still has to select appropriate hyper-parameters for the employed algorithm.

Therefore, we adopt the alternative approach of translating the proximal splitting algorithm into a neural network architecture by unrolling it for a fixed number of iterations K. Within every iteration the network alternates between processing the current estimate with a CNN, which replaces the proximal mapping on $\mathcal{R}$ in (4a), and enforcing data consistency as in (6). Concerning the latter, the scaling factors w_k are defined as learnable parameters in order to render the network end-to-end trainable. Parametrizing distinct CNNs for every iteration as in [4] and [19], however, would be redundant leading to increased memory requirements. Therefore, we propose to use a convolutional gated recurrent network (CGRN) which combines the benefits of weight-sharing across iterations with the ability of adapting to iteration-dependent changes (e.g. consider μ_k in (4a)) using an internal memory (see Fig. 1). The CGRN consists of U convolutional gated recurrent units [2] with a residual connection from the input to the output of the last unit. For a given unit indexed by $u = 1, ..., U$, the number of channels of the memory variable is represented by G_u.

Instead of operating on averaged DW images, we aim to exploit the correlations across the multiple acquisitions for every image slice by stacking them along an additional dimension interpreted as "image depth". Accordingly, 3-D convolutions are used within the units of the CGRN allowing to simultaneously extract features along depth and spatial dimensions. To accommodate for the complex nature of MR images, real and imaginary components are stacked along the channel dimension. As a result, the proposed network operates on fifth-order tensors with dimensions $B \times C \times D \times H \times W$ denoting batch, channel, depth, height and width dimension, respectively.

3   Evaluation

3.1   Data

Two prototype sequences – conventional single-shot EPI (SS-EPI, parallel acceleration factor of 2) and reduced field-of-view EPI (ZOOMit, no acceleration, phase field-of-view 50%) – were used to acquire DW images of the pelvis in 20 healthy male volunteers on various 1.5T and 3T MR scanners (MAGNETOM, Siemens Healthcare, Erlangen, Germany). Since a single-channel model is employed in this work, we operated on already phase-corrected and coil-combined data. In the case of conventional SS-EPI, fully-sampled data was generated by means of GRAPPA [6]. Acquisitions were performed with two common b-values ($b_{50} = 50\,\text{s/mm}^2$ and $b_{800} = 800\,\text{s/mm}^2$) as well as two different phase-encoding directions (RL and AP). The images were divided into training, validation and test set comprising the data of 16 (10,576 images), 2 and 2 volunteers (1,536 images each), respectively. For the volunteers assigned to the test set, prospectively sub-sampled data was acquired as well.

Images were pre-processed by normalizing the intensity range to $[-1, 1]$ and adding small complex Gaussian noise in order to train the regularizer to denoise as well. The standard deviation of the added noise was consistent with the acquired noise map (see supplementary material for details). For training, fully-sampled images underwent retrospective sub-sampling in addition.

3.2   Network Training

End-to-end training of the network was performed for a maximum of 100 epochs with early stopping. While the convolutional weights and biases were initialized using He initialization [9], parameters w_k were initialized with 1. All network parameters were optimized by means of Adam [11] with a learning rate of 10^{-4}

$(\beta_1 = 0.9, \beta_2 = 0.999)$ and a batch size of 1. The employed loss function $\mathcal{L}$ consisted of two terms weighted against each other by hyper-parameter τ:

$$\mathcal{L} = \mathcal{L}_1 + \tau \cdot \mathcal{L}_{\text{LPIPS}} . \tag{7}$$

Where as $\mathcal{L}_1$ denotes the L_1-penalized difference between the network prediction and the ground-truth averaged over all pixels, $\mathcal{L}_{\text{LPIPS}}$ is a neural network-based perceptual metric named *Learned Perceptual Image Patch Similarity* (LPIPS) [21] which was found to be sensitive with respect to blurring.

3.3 Experiments

In order to validate the proposed method, referred to as *Deep Recurrent Partial Fourier Network* (DRPF-Net) in the following, it was compared to the conventional POCS (5 iterations) as well as to a U-Net which models a direct image-to-image mapping. Note that we focus on the extreme case of a PF factor of 5/8 here. Three metrics were used for the quantitative evaluation: peak signal-to-noise ratio (PSNR), structural similarity (SSIM) and LPIPS. Further, the effect of PF-induced TE minimization on the quality of DW images acquired by SS-EPI was examined by comparing a fully-sampled reconstruction to the reconstructions of POCS and DRPF-Net on prospectively sub-sampled data. For all evaluations, the averaged, trace-weighted images were considered.

We set $\tau = 0.1$ in (7). Following hyper-parameter choices were made with respect to the proposed network architecture: $K = 5$, $U = 10$, $G_U = 2$ and $G_u = 64$ for $u = 1, ..., U-1$. Convolution kernels were of size $3 \times 3 \times 3$ and used zero-padding to keep the image size constant throughout the network. Please refer to the material for details on the implementation of the U-Net. The training scheme and objective were the same as described in Sect. 3.2.

Table 1. Mean PSNR, SSIM and LPIPS on the test set. Results are presented separately for the two employed b-values. Best and second-best results per metric are shown in bold and italic, respectively.

	b_{50}			b_{800}		
	PSNR (dB)	SSIM	LPIPS	PSNR (dB)	SSIM	LPIPS
ZF	31.83	.8912	.0765	31.25	.8751	.0822
POCS	32.78	.8964	.0338	32.29	.8828	.0341
U-Net	*33.65*	*.9163*	*.0218*	*32.48*	*.8906*	*.0321*
DRPF-Net	**35.22**	**.9358**	**.0155**	**33.54**	**.9077**	**.0236**

3.4 Results

Table 1 presents the mean values of the quantitative metrics achieved by the different reconstruction methods and zero-filling (ZF). DRPF-Net outperformed the other methods by relatively large margins across all three metrics. The U-Net

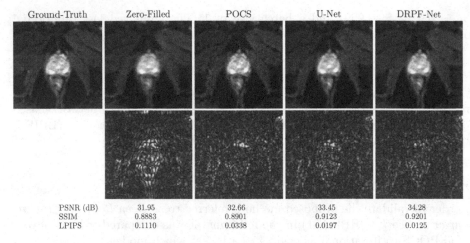

(a) Qualitative comparison of $b50$ images.

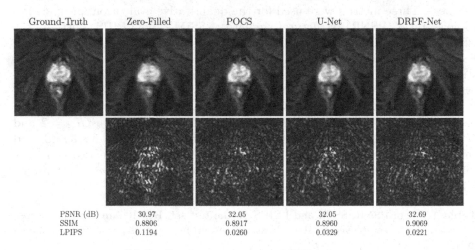

(b) Qualitative comparison of $b800$ images.

Fig. 2. Visual comparison of a representative prostate image acquired with SS-EPI at both b_{50} (a) and b_{800} (b). The bottom of each subfigure shows the residual maps with respect to the ground-truth.

is able to outperform POCS although the improvements are substantially smaller for $b800$ images than for $b50$ images. These observations were confirmed by the qualitative comparison of the reconstruction results provided in Fig. 2. While the reconstruction produced by ZF was clearly blurred along the RL direction, all other methods were able to restore image resolution to some extent. However, the noise added during pre-processing appeared to be amplified by POCS. The U-Net managed to reduce noise but noticeable residual blurring remained. In contrast, the difference maps of the proposed method exhibited less structure and noise than those of the reference methods.

Fully-sampled   POCS   DRPF-Net   DRPF-Net$_{NEX/2}$

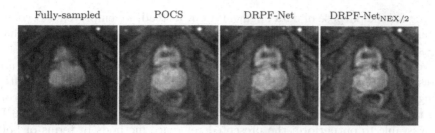

Fig. 3. SNR characteristics of a prostate image acquired with SS-EPI at $b800$ using fully-sampled and prospectively sub-sampled k-space where the latter was reconstructed by POCS and the proposed network using all (DRPF-Net) and half (DRPF-Net$_{NEX/2}$) of the excitations, respectively. Note that the fully-sampled and sub-sampled acquisitions did not exhibit the exact same anatomic structure due to subject motion in between them. The same windowing was used for all images in order to show the relative signal strengths.

The SNR gain resulting from PF sampling is demonstrated in Fig. 3. The fully-sampled acquisition had significantly lower signal compared to the reconstruction of prospectively sub-sampled raw data by means of POCS. However, the SNR gain was compromised by noise enhancement which was most apparent in the muscle tissue surrounding the prostate. In contrast, without sacrificing image sharpness, the reconstruction achieved by DRPF-Net was significantly smoother in regions which are supposed to be iso-intense. In fact, Fig. 3 also shows that when using only half the NEX, the proposed unrolled recurrent network still achieved SNR comparable to POCS.

4   Discussion and Conclusion

The conventional POCS performed poorly in the presence of noise and was limited by the regularization based on smoothness assumptions of the phase. Although the U-Net used almost twice the number of parameters compared to the employed DRPF-Net (10,069,346 vs. 5,662,986), it provided inferior results which may be due to the lack of data consistency considerations. In addition, the U-Net had to blindly adjust to images with significantly varying SNR. The proposed DRPF-Net alleviated those problems by employing data consistency operations within its architecture that allow to inject information on the physics-dependent noise level of a particular image.

Note that using convolutional recurrent units in unrolled network architectures was already proposed in [17] with application to dynamic MR reconstruction. However, the authors used basic Elman cells [5] with memory propagated bidirectionally: 1) along iterations and 2) along the time dimension of their data. Applying this idea to our problem in the sense that memory is additionally propagated along the dimension containing the averages would be less effective as the averages had no sequential context – in fact, they even were permuted during training for the purpose of data augmentation. Instead, we employed 3-D convolutions to process the additional dimension and used gated recurrent units which have superior memory conservation properties compared to Elman cells.

We presented an unrolled recurrent network which produces deblurred and denoised reconstructions of DW images acquired with PF sub-sampling. Alternating between a regularizing CGRN and data consistency operations that consider the spatially variant noise level of a particular image, the proposed DRPF-Net was able to outperform the presented reference methods in terms of both quantitative metrics as well as visual impression. It enables to realize DW images with significantly improved SNR or, alternatively, to reduce the NEX while achieving similar SNR as with fully-sampled acquisitions or POCS reconstructions. The extension of this method to multi-channel reconstruction will be the subject of our future research.

References

1. Aggarwal, H.K., Mani, M.P., Jacob, M.: MoDL: model-based deep learning architecture for inverse problems. IEEE Trans. Med. Imaging **38**(2), 394–405 (2018)
2. Ballas, N., Yao, L., Pal, C., Courville, A.: Delving deeper into convolutional networks for learning video representations. arXiv preprint arXiv:1511.06432 (2015)
3. Deng, J., Dong, W., Socher, R., Li, L.J., Li, K., Fei-Fei, L.: ImageNet: a large-scale hierarchical image database. In: 2009 IEEE Conference on Computer Vision and Pattern Recognition, pp. 248–255. IEEE (2009)
4. Diamond, S., Sitzmann, V., Heide, F., Wetzstein, G.: Unrolled optimization with deep priors. arXiv preprint arXiv:1705.08041 (2017)
5. Elman, J.L.: Finding structure in time. Cogn. Sci. **14**(2), 179–211 (1990)
6. Griswold, M.A., et al.: Generalized autocalibrating partially parallel acquisitions (GRAPPA). Magn. Reson. Med. Official J. Int. Soc. Magn. Reson. Med. **47**(6), 1202–1210 (2002)
7. Gupta, H., Jin, K.H., Nguyen, H.Q., McCann, M.T., Unser, M.: CNN-based projected gradient descent for consistent CT image reconstruction. IEEE Trans. Med. Imaging **37**(6), 1440–1453 (2018)
8. Haacke, E., Lindskogj, E., Lin, W.: A fast, iterative, partial-Fourier technique capable of local phase recovery. J. Magn. Reson. (1969) **92**(1), 126–145 (1991)
9. He, K., Zhang, X., Ren, S., Sun, J.: Delving deep into rectifiers: surpassing human-level performance on imagenet classification. In: Proceedings of the IEEE International Conference on Computer Vision, pp. 1026–1034 (2015)
10. Kellman, P., McVeigh, E.R.: Image reconstruction in SNR units: a general method for SNR measurement. Magn. Reson. Med. **54**(6), 1439–1447 (2005)
11. Kingma, D.P., Ba, J.: Adam: a method for stochastic optimization. arXiv preprint arXiv:1412.6980 (2014)
12. Maier, A., et al.: Precision learning: towards use of known operators in neural networks. In: 2018 24th International Conference on Pattern Recognition (ICPR), pp. 183–188. IEEE (2018)
13. Meinhardt, T., Moller, M., Hazirbas, C., Cremers, D.: Learning proximal operators: using denoising networks for regularizing inverse imaging problems. In: Proceedings of the IEEE International Conference on Computer Vision, pp. 1781–1790 (2017)

14. Mitchell, D.G., Bruix, J., Sherman, M., Sirlin, C.B.: LI-RADS (Liver Imaging Reporting and Data System): summary, discussion, and consensus of the LI-RADS Management Working Group and future directions. Hepatology **61**(3), 1056–1065 (2015)
15. Muckley, M.J., et al.: Training a neural network for Gibbs and noise removal in diffusion MRI. arXiv preprint arXiv:1905.04176 (2019)
16. Noll, D.C., Nishimura, D.G., Macovski, A.: Homodyne detection in magnetic resonance imaging. IEEE Trans. Med. Imaging **10**(2), 154–163 (1991)
17. Qin, C., Schlemper, J., Caballero, J., Price, A.N., Hajnal, J.V., Rueckert, D.: Convolutional recurrent neural networks for dynamic MR image reconstruction. IEEE Trans. Med. Imaging **38**(1), 280–290 (2018)
18. Ronneberger, O., Fischer, P., Brox, T.: U-Net: convolutional networks for biomedical image segmentation. In: Navab, N., Hornegger, J., Wells, W.M., Frangi, A.F. (eds.) MICCAI 2015. LNCS, vol. 9351, pp. 234–241. Springer, Cham (2015). https://doi.org/10.1007/978-3-319-24574-4_28
19. Schlemper, J., Caballero, J., Hajnal, J.V., Price, A., Rueckert, D.: A deep cascade of convolutional neural networks for MR image reconstruction. In: Niethammer, M., et al. (eds.) IPMI 2017. LNCS, vol. 10265, pp. 647–658. Springer, Cham (2017). https://doi.org/10.1007/978-3-319-59050-9_51
20. Turkbey, B., et al.: Prostate imaging reporting and data system version 2.1: 2019 update of prostate imaging reporting and data system version 2. European urology (2019)
21. Zhang, R., Isola, P., Efros, A.A., Shechtman, E., Wang, O.: The unreasonable effectiveness of deep features as a perceptual metric. In: Proceedings of the IEEE Conference on Computer Vision and Pattern Recognition, pp. 586–595 (2018)

Model-Based Learning for Quantitative Susceptibility Mapping

Juan Liu[1,2(✉)] and Kevin M. Koch[1,2,3]

[1] Center for Imaging Research, Medical College of Wisconsin, Milwaukee, WI, USA
[2] Department of Biomedical Engineering, Marquette University and Medical College of Wisconsin, Milwaukee, WI, USA
juan.liu@marquette.edu
[3] Department of Radiology, Medical College of Wisconsin, Milwaukee, WI, USA

Abstract. Quantitative susceptibility mapping (QSM) is a magnetic resonance imaging (MRI) technique that estimates magnetic susceptibility of tissue from Larmor frequency offset measurements. The generation of QSM requires solving a challenging ill-posed field-to-source inversion problem. Inaccurate field-to-source inversion often causes large susceptibility estimation errors that appear as streaking artifacts in the QSM, especially in massive hemorrhagic regions. Recently, several deep learning (DL) QSM techniques have been proposed and demonstrated impressive performance. Due to the inherent non-existent ground-truth QSM references, these DL techniques used either calculation of susceptibility through multiple orientation sampling (COSMOS) maps or synthetic data for network training. Therefore, they were constrained by the availability and accuracy of COSMOS maps, or suffered from performance drop when the training and testing domains were different. To address these limitations, we present a model-based DL method, denoted as uQSM. Without accessing to QSM labels, uQSM is trained using the well-established physical model. When evaluating on multi-orientation QSM datasets, uQSM achieves higher levels of quantitative accuracy compared to TKD, TV-FANSI, MEDI, and DIP approaches. When qualitatively evaluated on single-orientation datasets, uQSM outperforms other methods and reconstructed high quality QSM.

Keywords: Quantitative susceptibility mapping · Self-supervised learning · Dipole inversion

1 Introduction

Quantitative susceptibility mapping (QSM) can estimate tissue magnetic susceptibility values from magnetic resonance imaging (MRI) Larmor frequency

Electronic supplementary material The online version of this chapter (https://doi.org/10.1007/978-3-030-61598-7_5) contains supplementary material, which is available to authorized users.

sensitive phase images [31]. Biological tissue magnetism can provide useful diagnostic image contrast and be used to quantify biomarkers including iron, calcium, and gadolinium [31]. To date, all QSM methods rely on a dipolar convolution that relates susceptibility sources to induced Larmor frequency offsets [20,25], which is expressed in the k-space as bellow.

$$B(\boldsymbol{k}) = X(\boldsymbol{k}) \cdot D(\boldsymbol{k}); D(\boldsymbol{k}) = \frac{1}{3} - \frac{k_z^2}{k_x^2 + k_y^2 + k_z^2} \tag{1}$$

where $B(\boldsymbol{k})$ is the susceptibility induced magnetic perturbation along the main magnetic field direction, $X(\boldsymbol{k})$ is the susceptibility distribution χ in the k space, $D(\boldsymbol{k})$ is the dipole kernel. While the forward relationship of this model (source to field) is well-established and can be efficiently computed using Fast-Fourier-Transform (FFT), the k-space singularity in the dipole kernel results in an ill-conditioned relationship in the field-to-source inversion.

Calculation of susceptibility through multiple orientation sampling (COSMOS) [18] remains the empirical gold-standard of QSM, as the additional field data sufficiently improves the conditioning of the inversion algorithm. Since it is time-consuming and clinically infeasible to acquire multi-orientation data, single-orientation QSM is preferred which is computed by either thresholding of the convolution operator [9,27,32] or use of more sophisticated regularization methods [2,6,17,24]. In single-orientation QSM, inaccurate field-to-source inversion often causes large susceptibility estimation errors that appear as streaking artifacts in the QSM, especially in massive hemorrhagic regions.

Recently, several deep learning (DL) approaches have been proposed to solve for the QSM dipole inversion. QSMnet [34] used COSMOS results as QSM labels for training, which reconstructed COSMOS-like QSM estimates no matter the head orientations. DeepQSM [3] used synthetic susceptibility maps simulated using basic 3D geometric shapes and the forward dipole model to generate synthetic training data. QSMGAN [5] adopted COSMOS maps as QSM labels and refined the network using the Wasserstein Generative Adversarial Networks (WGAN) [1,7]. QSMnet+ [12] employed data augmentation approaches to increase the range of susceptibility, which improved the linearity of susceptibility measurement in clinical situations.

Though these DL techniques have exhibited impressive results, there were several limitations. These methods are supervised and data-driven which require QSM labels for network training. Unfortunately, QSM has the inherent non-existent 'ground-truth'. Therefore, these methods used either COSMOS data or synthetic data for network training. However, acquiring a large number of COSMOS data is not only expensive but also time consuming. Moreover, COSMOS neglects tissue susceptibility anisotropy [15] and contains errors from background field removal and image registration procedures, which compromises COSMOS map as a QSM label. Though synthetic data provides a reliable and cost-effective way for training, the generalization capability needs to be addressed since the domain gap between the synthetic training data and real data often causes performance degradation and susceptibility quantification errors.

Here, we propose a model-based learning method without the need of QSM labels for QSM dipole inversion, denoted as uQSM, to overcome these limitations. Quantitative evaluation is performed on multi-orientation datasets in comparison to TKD [27], TV-FANSI [21], MEDI [16], and deep image prior (DIP) [30], with COSMOS result as a reference. In addition, qualitative evaluation is performed on single-orientation datasets.

2   Method

uQSM adopted a 3D convolutional neural network with an encoder-decoder architecture as shown in Fig. 1.

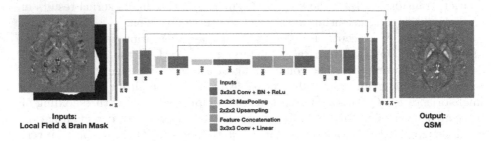

Fig. 1. Neural network architecture of uQSM. It has an encoder-decoder structure with 9 convolutional layers (kernel size $3 \times 3 \times 3$, same padding), 9 batch normalization layers, 9 ReLU layers, 4 max pooling layers (pooling size $2 \times 2 \times 2$, strides $2 \times 2 \times 2$), 4 nearest-neighbor upsampling layers (size $2 \times 2 \times 2$), 4 feature concatenations, and 1 convolutional layer (kernel size $3 \times 3 \times 3$, linear activation).

The network took two inputs, a local field measurement f and a brain mask m, and one output, a susceptibility map χ. Upsampling layers were used in the decoding path in stead of deconvolutional layers to address the checkboard artifacts [22] in the reconstructed QSM.

The loss function incorporated the model-based data consistency loss L_χ.

$$L_\chi = \left\| Wm(e^{jd*\chi} - e^{jf}) \right\|_2 \tag{2}$$

where W serves as a data-weighting factor which can be the magnitude image or noise weight matrix, d is the dipole kernel, $*$ is the convolution operator. Since noise is unknown and spatially variant in the local field measurements, the nonlinear dipole convolution data consistency loss was used in the loss function to get more robust QSM estimates as conventional QSM methods [19, 23]. The dipole convolution was computed in the k-space using FFT.

For normalization, the mean and standard deviation were calculated in the local fields. Then, the input local maps were normalized to have a mean of 0 and a standard deviation of 1. Since the susceptibility maps were unknown, we used

$3d * \chi$ to be consistent with f, which can make the susceptibility outputs close to being standard normalized.

$$L_{TV} = \|G_x(\chi)\|_1 + \|G_y(\chi)\|_1 + \|G_z(\chi)\|_1 \tag{3}$$

In addition, a total variation (TV) loss L_{TV} was included to serve as a regularization term to preserve important details such as edges whilst removing unwanted noise in the reconstructed susceptibility maps. In L_{TV}, G_x, G_y, G_z are gradient operators in x, y, z directions.

$$L_{Total} = L_\chi + \lambda L_{TV} \tag{4}$$

The loss function is the weighted sum of the data consistency loss L_χ and the total variation loss L_{TV}.

3   Experiments

Multi-orientation QSM Data. 9 QSM datasets were acquired using 5 head orientations and a 3D single-echo GRE scan with isotropic voxel size $1.0 \times 1.0 \times 1.0 \, \text{mm}^3$ on 3T MRI scanners. QSM data processing was implemented as following, offline GRAPPA [8] reconstruction to get magnitude and phase images from saved k-space data, coil combination using sensitivities estimated with ESPIRiT [29], BET (FSL, FMRIB, Oxford, UK) [28] for brain extraction, Laplacian method [14] for phase unwrapping, and RESHARP [33] with spherical mean radius 4mm for background field removal. COSMOS results were calculated using the 5 head orientation data which were registered by FLIRT (FSL, FMRIB, Oxford, UK) [10,11]. In addition, QSM estimates at the normal head position were generated using the TKD, TV-FANSI, and MEDI algorithms.

For uQSM, leave-one-out cross validation was used. For each dataset, total 40 scans from other 8 datasets were used for training. uQSM was trained using patch-based neural network with patch size $96 \times 96 \times 96$. The RESHARP local field and brain mask patches with patch size $96 \times 96 \times 96$ were cropped with an overlapping scheme of 16.6% overlap between adjacent patches. These patch pairs were used for training and validation with split ratio 9:1. The magnitude images were scaled between 0 to 1 and used as the weighting factor W, λ was set 0.001. The Adam optimizer [13] was used for the model training. The initial learning rate was set as 0.0001, with exponentially decay at every 100 steps. One NVIDIA GPU Tesla k40 was used for training with batch size 4. The model was trained and evaluated using Keras with Tensorflow as a backend. After training, the full local field and brain mask from the leave-one dataset were fed to the trained model to get the QSM estimates.

In addition, we used DIP to get the QSM images. DIP used the same neural network architecture and loss function described above. DIP was performed on each individual dataset using full neural network. To avoid overfitting which can introduce artifacts in the reconstructed QSM images, DIP was stopped after 200 iterations to get QSM results.

The QSM of uQSM, TKD, TV-FANSI, MEDI, and DIP were compared with respect to the COSMOS maps using quantitative metrics, peak signal-to-noise ratio (pSNR), normalized root mean squared error (NRMSE), high frequency error norm (HFEN), and structure similarity (SSIM) index.

Single-orientation QSM Data. 150 QSM datasets were collected on a 3T MRI scanner (GE Healthcare MR750) from a commercially available susceptibility weighted software application (SWAN, GE Healthcare). The data acquisition parameters were as follows: in-plane data matrix 320×256, field of view 24 cm, voxel size $0.5 \times 0.5 \times 2.0\,\mathrm{mm}^3$, 4 echo times [10.4, 17.4, 24.4, 31.4] ms, repetition time 58.6 ms, autocalibrated parallel imaging factors 3×1, and total acquisition time 4 min.

Complex multi-echo images were reconstructed from raw k-space data using customized code. The brain masks were obtained using the SPM tool [4]. After background field removal using the RESHARP with spherical mean radius 4mm, susceptibility inversion was performed using TKD, TV-FANSI, and MEDI. In addition, we used DIP to get the QSM images. DIP was performed for each individual dataset and early stopped after 200 iterations.

For uQSM training, 10449 patch pairs of local field maps and brain masks with patch size $128 \times 128 \times 64$ were extracted from 100 QSM datasets. The average of multi-echo magnitude images was scaled between 0 to 1 and used as the weighting factor W. In the loss function, λ was set 0.0 since the data had high signal-to-noise ratio. The Adam optimizer was used with an initial learning rate 0.0001, which exponentially decayed at every 100 steps. One NVIDIA GPU Tesla k40 was used for training with batch size 2. After training, the trained DL model took full local fields and brain masks to get the QSM estimates.

4   Experimental Results

Table 1. Means and standard deviations of quantitative performance metrics of 5 reconstructed QSM images with COSMOS as a reference on 9 multi-orientation datasets.

	pSNR (dB)	NRMSE (%)	HFEN (%)	SSIM (0-1)
TKD	43.4 ± 0.5	91.4 ± 6.7	72.9 ± 6.6	0.831 ± 0.016
TV-FANSI	41.5 ± 0.6	80.0 ± 5.0	73.6 ± 6.2	0.869 ± 0.019
MEDI	41.5 ± 0.6	113.8 ± 7.6	100.4 ± 9.1	**0.902 ± 0.016**
DIP	44.0 ± 0.8	85.5 ± 6.7	65.7 ± 4.5	0.859 ± 0.020
uQSM	**45.6 ± 0.4**	**71.4 ± 5.0**	**62.8 ± 5.0**	0.890 ± 0.015

Multi-orientation QSM Data. Table 1 summarized quantitative metrics of 5 reconstruction methods on 9 multi-orientation datasets with COSMOS map as a reference. Compared to TKD, TV-FANSI, MEDI, and DIP, uQSM results achieved the best metric scores in pSNR, RMSE, and HFEN, and second in SSIM. Figure 2 compared QSM images from a representative dataset in three planes. Streaking artifacts were observed in the sagittal planes of TKD, TV-FANSI, and MEDI results (a-c, iii, black solid arrows). TV-FANSI and MEDI maps showed substantial blurring due to their use of spatial regularization. DIP results displayed good image quality. uQSM demonstrated superior image sharpness and invisible image artifacts. Compared with uQSM, COSMOS results displayed conspicuity loss due to image registration errors (e–f, i, black dash arrows).

Single-orientation QSM Data. Figure 3 displayed QSM images from a single-orientation dataset. TKD, MEDI results had black shading artifacts in the axial plane, and streaking artifacts in coronal and sagittal planes. MEDI and TV-FANSI images showed oversmoothing and lost image sharpness. DIP results showed high quality but subtle image artifacts. Visual comparison demonstrated that uQSM outperformed other methods and produced better QSM images.

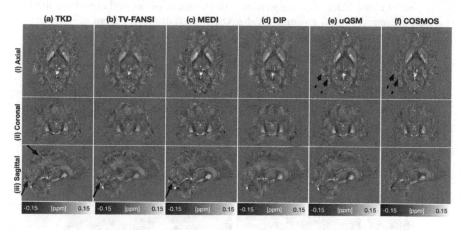

Fig. 2. Comparison of QSM of a multi-orientation data. TKD (a), TV-FANSI (b) and MEDI (c) maps showed oversmoothing and/or streaking artifacts. The uQSM (e) maps well preserve image details and show invisible artifacts.

Deconvolution and Checkboard Artifacts. In uQSM, we used upsampling layers rather than deconvolutional layers to reduce the checkerboard artifacts [22] in QSM results. Here we trained two networks on the single-orientation datasets, one with upsampling and the other with deconvolution in the decoding path. Figure 4 compared QSM images of 2 single-orientation datasets reconstructed using deconvolution-based network and upsampling-based network. Deconvolution-based network produced QSM with checkboard artifacts in the zoom-in axial plane (a, ii, black arrows).

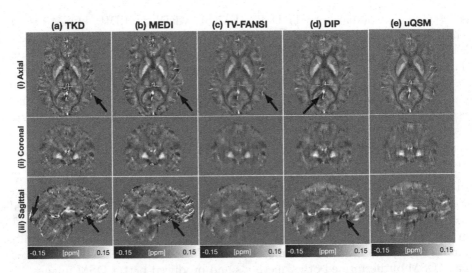

Fig. 3. Comparison of QSM of a single-orientation dataset. TDK (a), MEDI (b), and DIP (d) results show black shading artifacts in the axial plane and streaking artifacts in the sagittal plane. MEDI (b) and TV-FANSI (c) results suffer from oversmoothing. uQSM (e) images have high-quality with clear details and invisible artifacts.

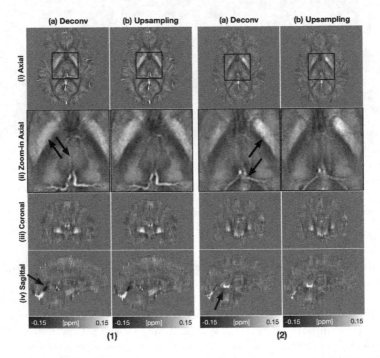

Fig. 4. Comparison of QSM results of 2 single-orientation datasets. uQSM using deconvolution (a) shows checkboard artifacts in zoom-in axial plane (a, ii, black arrows).

Effects of L_{TV} and Comparison of Data Consistency Losses. We used the multi-orientation datasets to investigate the effects of L_{TV} and three data consistency losses - (1) linear dipole inversion (LDI), $L_{LDI} = \|m(d * \chi - y)\|_2$, (2) weighted linear dipole inversion (WLDI), $L_{WLDI} = \|Wm(e^{id*\chi} - e^{iy})\|_2$, (3) weighted nonlinear dipole inversion (NDI), $L_{NDI} = \|Wm(d * \chi - y)\|_2$.

Table 2. Means and standard deviations of quantitative performance metrics of uQSM using different loss function on 9 multi-orientation datasets.

	pSNR (dB)	NRMSE (%)	HFEN (%)	SSIM (0-1)
L_{NDI}	43.8 ± 0.5	87.4 ± 6.8	70.5 ± 5.7	0.848 ± 0.022
$L_{LDI} + \lambda L_{TV}$	44.1 ± 0.5	84.9 ± 5.9	73.4 ± 5.7	0.879 ± 0.013
$L_{LWDI} + \lambda L_{TV}$	45.0 ± 0.5	75.9 ± 5.4	67.1 ± 5.5	0.888 ± 0.015
$L_{NDI} + \lambda L_{TV}$	$\mathbf{45.6 \pm 0.4}$	$\mathbf{71.4 \pm 5.0}$	$\mathbf{62.8 \pm 5.0}$	$\mathbf{0.890 \pm 0.015}$

Table 2 summarized quantitative metrics on 9 multi-orientation datasets with COSMOS map as a reference. uQSM using $L_{NDI} + \lambda L_{TV}$ achieved the best metric scores. Figure 5 displayed QSM images. Without L_{TV}, the QSM showed high level of noise (a). Using L_{LDI} and L_{WLDI} as data consistency loss, the QSM estimates displayed black shading artifacts (b–c, i–iii, black arrows), while L_{NDI} was capable of suppressing these artifacts.

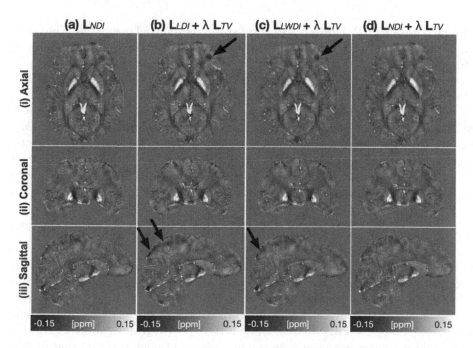

Fig. 5. Comparison of uQSM results using different loss functions of one multi-orientation datasets. Without L_{TV}, QSM shows high level of noise (a). Using L_{LDI} and L_{WLDI} as data consistency loss, the QSM estimates display black shading artifacts (b–c, i–iii, black arrows), while L_{NDI} suppresses these artifacts.

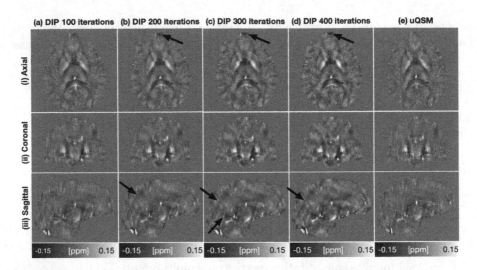

Fig. 6. Comparison of DIP results at 100, 200, 300, and 400 iterations from a single-orientation dataset. DIP results of 100 iterations (a) show inferior image contrast yet invisible image artifacts. Streaking and shading artifacts become more apparent in DIP results of 200, 300, and 400 iterations (b–d, black arrows). Visual assessment demonstrates that uQSM (e) reconstructs QSM with better quality and invisible artifacts.

Overfitting of DIP. DIP requires early-stopping to avoid overfitting which can result in artifacts in the reconstructed QSM images. In this paper, DIP was early stopping after 200 iterations to get QSM estimates. In order to better understand the overfitting problem in DIP, here we displayed the DIP results at 100, 200, 300, and 400 iterations from a single-orientation datasets.

Figure 6 displayed the DIP results at 100, 200, 300, and 400 iterations, and uQSM images from a multi-orientation dataset. It was clearly observed than DIP results at 100 iterations displayed inferior image contrast but less image artifacts. DIP results at 200 iterations results displaying high quality QSM images with subtle artifacts, while DIP results at 300 and 400 iterations results showed severe shading streaking artifacts due to overfitting.

5   Discussion and Conclusion

In this work, a model-based DL method for QSM dipole deconvolution was proposed. Without accessing to QSM labels during training, uQSM learned to perform dipole inversion through the physical model.

From quantitative evaluation on multi-orientation QSM datasets, uQSM outperformed TKD, TV-FANSI, MEDI, and DIP in pSNR, RMSE, and HFEN, with COSMOS map as a reference. The visual assessment demonstrated that uQSM preserved image details well and showed invisible image artifacts. When using single-orientation datasets for qualitative assessment, uQSM results showed better image quality than conventional non-DL methods and DIP. Though DIP is unsupervised and does not require training data for prior training, it needs long iteration times for each dataset and early-stopping to avoid overfitting. In addition, the upsampling layers used in uQSM network can avoid the checkerboard artifacts in QSM estimates. L_{TV} enables to denosing the QSM outputs and preserves the edge information. L_{NDI} as data consistency loss can improve the image quality of uQSM than L_{LDI} and L_{LWDI}.

Future work can apply more sophisticated models [26] in uQSM. In addition, uQSM is still affected by the performance of background field removal methods. It is necessary to investigate the effects of background field removal on susceptibility quantification or perform DL-based single-step QSM reconstruction.

Acknowledgement. We thank Professor Jongho Lee for sharing the multi-orientation QSM datasets.

References

1. Arjovsky, M., Chintala, S., Bottou, L.: Wasserstein generative adversarial networks. In: International Conference on Machine Learning, pp. 214–223 (2017)
2. Bilgic, B., Chatnuntawech, I., Fan, A.P., et al.: Fast image reconstruction with L2-regularization. J. Magn. Reson. Imaging **40**(1), 181–191 (2014)
3. Bollmann, S., Rasmussen, K.G.B., Kristensen, M., et al.: DeepQSM-using deep learning to solve the dipole inversion for quantitative susceptibility mapping. NeuroImage **195**, 373–383 (2019)
4. Brett, M., Anton, J.L., Valabregue, R., et al.: Region of interest analysis using an SPM toolbox. In: 8th International Conference on Functional Mapping of the Human Brain, Sendai, p. 497 (2002)
5. Chen, Y., Jakary, A., Avadiappan, S., et al.: QSMGAN: improved quantitative susceptibility mapping using 3D generative adversarial networks with increased receptive field. NeuroImage **207**, 116389 (2019)
6. De Rochefort, L., Brown, R., Prince, M.R., et al.: Quantitative MR susceptibility mapping using piece-wise constant regularized inversion of the magnetic field. Magn. Reson. Med. **60**(4), 1003–1009 (2008)
7. Goodfellow, I., Pouget-Abadie, J., Mirza, M., et al.: Generative adversarial nets. In: Advances in Neural Information Processing Systems, pp. 2672–2680 (2014)
8. Griswold, M.A., et al.: Generalized autocalibrating partially parallel acquisitions (GRAPPA). Magn. Reson. Med. **47**(6), 1202–1210 (2002)
9. Haacke, E., Tang, J., Neelavalli, J., et al.: Susceptibility mapping as a means to visualize veins and quantify oxygen saturation. J. Magn. Reson. Imaging **32**(3), 663–676 (2010)
10. Jenkinson, M., Bannister, P., Brady, M., et al.: Improved optimization for the robust and accurate linear registration and motion correction of brain images. Neuroimage **17**(2), 825–841 (2002)
11. Jenkinson, M., Smith, S.: A global optimisation method for robust affine registration of brain images. Med. Image Anal. **5**(2), 143–156 (2001)
12. Jung, W., Yoon, J., Ji, S., et al.: Exploring linearity of deep neural network trained QSM: QSMnet+. NeuroImage **211**, 116619 (2020)

13. Kingma, D.P., Ba, J.: Adam: a method for stochastic optimization. arXiv (2014)
14. Li, W., Wu, B., Liu, C.: Quantitative susceptibility mapping of human brain reflects spatial variation in tissue composition. NeuroImage **55**(4), 1645–1656 (2011)
15. Liu, C.: Susceptibility tensor imaging. Magn. Reson. Med. **63**(6), 1471–1477 (2010)
16. Liu, J., Liu, T., de Rochefort, L., et al.: Morphology enabled dipole inversion for quantitative susceptibility mapping using structural consistency between the magnitude image and the susceptibility map. NeuroImage **59**(3), 2560–2568 (2012)
17. Liu, T., Liu, J., De Rochefort, L., et al.: Morphology enabled dipole inversion (MEDI) from a single-angle acquisition: comparison with COSMOS in human brain imaging. Magn. Reson. Med. **66**(3), 777–783 (2011)
18. Liu, T., Spincemaille, P., De Rochefort, L., et al.: Calculation of susceptibility through multiple orientation sampling (COSMOS): a method for conditioning the inverse problem from measured magnetic field map to susceptibility source image in MRI. Magn. Reson. Med. **61**(1), 196–204 (2009)
19. Liu, T., Wisnieff, C., Lou, M., et al.: Nonlinear formulation of the magnetic field to source relationship for robust quantitative susceptibility mapping. Magn. Reson. Med. **69**(2), 467–476 (2013)
20. Marques, J., Bowtell, R.: Application of a Fourier-based method for rapid calculation of field inhomogeneity due to spatial variation of magnetic susceptibility. Concepts Magn. Reson. Part B Magn. Reson. Eng. **25**(1), 65–78 (2005)
21. Milovic, C., Bilgic, B., Zhao, B., et al.: Fast nonlinear susceptibility inversion with variational regularization. Magn. Reson. Med. **80**(2), 814–821 (2018)
22. Odena, A., Dumoulin, V., Olah, C.: Deconvolution and checkerboard artifacts. Distill **1**(10), e3 (2016)
23. Polak, D., Chatnuntawech, I., Yoon, J., et al.: Nonlinear dipole inversion (NDI) enables robust quantitative susceptibility mapping (QSM). NMR Biomed., e4271 (2020)
24. de Rochefort, L., Liu, T., Kressler, B., et al.: Quantitative susceptibility map reconstruction from MR phase data using Bayesian regularization: validation and application to brain imaging. Magn. Reson. Med. **63**(1), 194–206 (2010)
25. Salomir, R., de Senneville, B.D., Moonen, C.T.: A fast calculation method for magnetic field inhomogeneity due to an arbitrary distribution of bulk susceptibility. Concepts Magn. Reson. Part B Magn. Reson. Eng. **19**(1), 26–34 (2003)
26. Schweser, F., Zivadinov, R.: Quantitative susceptibility mapping (QSM) with an extended physical model for MRI frequency contrast in the brain: a proof-of-concept of quantitative susceptibility and residual (QUASAR) mapping. NMR Biomed. **31**(12), e3999 (2018)
27. Shmueli, K., de Zwart, J.A., van Gelderen, P., et al.: Magnetic susceptibility mapping of brain tissue in vivo using MRI phase data. Magn. Reson. Med. **62**(6), 1510–1522 (2009)
28. Smith, S.M.: Fast robust automated brain extraction. Hum. Brain Mapp. **17**(3), 143–155 (2002)
29. Uecker, M., Lai, P., Murphy, M.J., et al.: ESPIRiT-an eigenvalue approach to autocalibrating parallel MRI: where sense meets GRAPPA. Magn. Reson. Med. **71**(3), 990–1001 (2014)
30. Ulyanov, D., Vedaldi, A., Lempitsky, V.: Deep image prior. In: Proceedings of the IEEE CVPR, pp. 9446–9454 (2018)
31. Wang, Y., Liu, T.: Quantitative susceptibility mapping (QSM): decoding MRI data for a tissue magnetic biomarker. Magn. Reson. Med. **73**(1), 82–101 (2015)

32. Wharton, S., Schäfer, A., Bowtell, R.: Susceptibility mapping in the human brain using threshold-based k-space division. Magn. Reson. Med. **63**(5), 1292–1304 (2010)
33. Wu, B., Li, W., Guidon, A., Liu, C.: Whole brain susceptibility mapping using compressed sensing. Magn. Reson. Med. **67**(1), 137–147 (2012)
34. Yoon, J., Gong, E., Chatnuntawech, I., et al.: Quantitative susceptibility mapping using deep neural network: QSMnet. NeuroImage **179**, 199–206 (2018)

Learning Bloch Simulations for MR Fingerprinting by Invertible Neural Networks

Fabian Balsiger[1,2](✉) , Alain Jungo[1,2](✉) , Olivier Scheidegger[3], Benjamin Marty[4,5], and Mauricio Reyes[1,2]

[1] ARTORG Center for Biomedical Engineering Research, University of Bern, Bern, Switzerland
{fabian.balsiger,alain.jungo}@artorg.unibe.ch
[2] Insel Data Science Center, Inselspital, Bern University Hospital, Bern, Switzerland
[3] Support Center for Advanced Neuroimaging (SCAN), Institute for Diagnostic and Interventional Neuroradiology, Inselspital, Bern University Hospital, Bern, Switzerland
[4] NMR Laboratory, Institute of Myology, Neuromuscular Investigation Center, Paris, France
[5] NMR Laboratory, CEA, DRF, IBFJ, MIRCen, Paris, France

Abstract. Magnetic resonance fingerprinting (MRF) enables fast and multiparametric MR imaging. Despite fast acquisition, the state-of-the-art reconstruction of MRF based on dictionary matching is slow and lacks scalability. To overcome these limitations, neural network (NN) approaches estimating MR parameters from fingerprints have been proposed recently. Here, we revisit NN-based MRF reconstruction to jointly learn the forward process from MR parameters to fingerprints and the backward process from fingerprints to MR parameters by leveraging invertible neural networks (INNs). As a proof-of-concept, we perform various experiments showing the benefit of learning the forward process, i.e., the Bloch simulations, for improved MR parameter estimation. The benefit especially accentuates when MR parameter estimation is difficult due to MR physical restrictions. Therefore, INNs might be a feasible alternative to the current solely backward-based NNs for MRF reconstruction.

Keywords: Reconstruction · Magnetic resonance fingerprinting · Invertible neural network

1 Introduction

Magnetic resonance fingerprinting (MRF) [16] is a relatively new but increasingly used [20] concept for fast and multiparametric quantitative MR imaging.

F. Balsiger and A. Jungo—These authors contributed equally and are listed by flipping a coin.

F. Deeba et al. (Eds.): MLMIR 2020, LNCS 12450, pp. 60–69, 2020.
https://doi.org/10.1007/978-3-030-61598-7_6

Acquisitions of MRF produce unique magnetization evolutions per voxel, called fingerprints, due to temporal varying MR sequence schedules. From these fingerprints, MR parameters (e.g., relaxation times) are then reconstructed using a dictionary matching, comparing each fingerprint to a dictionary of simulated fingerprints with known MR parameters. Although the MRF acquisition itself is fast thanks to high undersampling, the dictionary matching is slow, discrete and cannot interpolate, and lacks scalability with increasing number of MR parameters.

With the advent of deep learning, neural networks (NNs) have been explored to overcome the limitations of the dictionary matching. The dictionary matching can be formulated as a regression problem from the fingerprints to the MR parameters. Several methods have been applied to MRF with impressive results both in terms of reconstruction accuracy and speed [3,4,8,10–14,19,21]. Among these, spatially regularizing methods trained on *in vivo* MRF acquisitions showed superiority over methods performing fingerprint-wise regression [3–5,10,11,14]. However, spatial methods might require a considerable amount of training data to achieve reasonable robustness for highly heterogeneous diseases [3]. Therefore, robust fingerprint-wise methods, leveraging the dictionaries for training, are required to alleviate the need of *in vivo* MRF acquisitions.

We revisit NN-based MRF reconstruction by formulating it as an inverse problem where we jointly learn the forward process from MR parameters to fingerprints and the backward process from fingerprints to MR parameters. In doing so, the available information of the forward process is leveraged, which might help disentangling MR physical processes and consequently improve the MR parameter estimation of the backward process. To this end, we leverage invertible neural networks (INNs) [9]. As proof-of-concept, we perform various experiments showing the benefit of learning the forward process, i.e., the Bloch simulations, for improved NN-based MRF reconstruction.

2   Methodology

2.1   MR Fingerprinting Using Invertible Neural Networks

Inverse problems are characterized by having some observations $\mathbf{y}$, from which we want to obtain the underlying parameters $\mathbf{x}$. The forward process $\mathbf{y} = f(\mathbf{x})$ is usually well defined and computable. However, the backward process $\mathbf{x} = f^{-1}(\mathbf{y})$ is not trivial to compute. MRF can be formulated as an inverse problem [7]. The forward process f is described by the Bloch equations [6]. Meaning, from some MR parameters $\mathbf{x} \in \mathbb{R}^M$, one can simulate a corresponding fingerprint $\mathbf{y} \in \mathbb{C}^T$ for a given MRF sequence. The backward process f^{-1} in MRF is typically solved by dictionary matching, or recently via regression by NNs. However, in doing so, the knowledge of the well-defined forward process is completely omitted in the backward process. We hypothesize that by leveraging the knowledge of the forward process, NN-based MRF reconstruction can be improved. Therefore, we aim at jointly learning the forward and the backward process by using INNs.

Once learned, the trained INN can be used to estimate MR parameters $\mathbf{x}$ from a fingerprint $\mathbf{y}$, as done in literature.

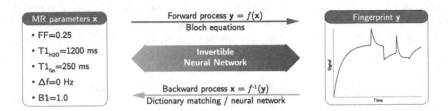

Fig. 1. Overview of the INN in the context of MRF. The forward process simulates fingerprints $\mathbf{y}$ from MR parameters $\mathbf{x}$, usually by Bloch simulations. The backward process estimates MR parameters $\mathbf{x}$ of a fingerprint $\mathbf{y}$, usually by dictionary matching or recently NNs. The INN is capable of doing both the forward and backward process.

Figure 1 depicts INNs in the context of MRF. Given training pairs $(\mathbf{x}, \mathbf{y})$ from a dictionary, the MR parameters $\mathbf{x}$ are fed into the INN, which predicts the fingerprint $\hat{\mathbf{y}}$. Optimizing a mean squared error (MSE) loss between $\mathbf{y}$ and $\hat{\mathbf{y}}$ results in learning the forward process. Feeding the fingerprint $\mathbf{y}$ from the opposite direction into the INN predicts the MR parameters $\hat{\mathbf{x}}$. Here, we also optimize a MSE loss between $\mathbf{x}$ and $\hat{\mathbf{x}}$ to learn the backward process[1]. For both forward and backward, the INN uses the same weights, and, therefore, the training jointly optimizes the forward and backward process.

The architecture of our INN bases on RealNVP [9] and consists of two reversible blocks with permutation layers [2]. A reversible block is composed of two complementary affine transformations, with scales s_i and translations t_i ($i \in \{1, 2\}$). The transformations describe the forward pass as

$$\mathbf{v}_1 = \mathbf{u}_1 \odot \exp(s_2(\mathbf{u}_2)) + t_2(\mathbf{u}_2), \quad \mathbf{v}_2 = \mathbf{u}_2 \odot \exp(s_1(\mathbf{v}_1)) + t_1(\mathbf{v}_1)$$

where $\mathbf{u} = [\mathbf{u}_1, \mathbf{u}_2]$ and $\mathbf{v} = [\mathbf{v}_1, \mathbf{v}_2]$ are the input and output split into halves, and $\odot$ is the Hadamard product. The reversibility of the affine transformations ensure the invertibility of the reversible block, such that the inverse is given by

$$\mathbf{u}_2 = (\mathbf{v}_2 - t_1(\mathbf{v}_1)) \odot \exp\left(-s_1(\mathbf{v}_1)\right), \quad \mathbf{u}_1 = (\mathbf{v}_1 - t_2(\mathbf{u}_2)) \odot \exp(-s_2(\mathbf{u}_2))$$

As a consequence, the operations s and t do not need to be invertible themselves. For each s_i and t_i, we use two fully-connected layers, with 128 neurons each, followed by ReLU and linear activation, respectively. The permutation layers enforce a different split of the halves in every reversible block [2]. We remark that we zero-pad the input $\mathbf{x}$ to match the dimensionality of $\mathbf{y}$. Generally, such INN architectures have been shown to be suitable to solve diverse inverse problems [2], including problems in medical imaging [1].

---

[1] The MSE loss was empirically found to be beneficial although the backward process is theoretically learned through the bijectivity property of the INN.

2.2   MR Fingerprinting Sequence

In the context of our clinical scope, we use MRF T1-FF [17], a MRF sequence designed for the quantification of T1 relaxation time (T1) and fat fraction (FF) in fatty infiltrated tissues such as diseased skeletal muscle. Since fat can heavily bias the T1 quantification, MRF T1-FF separately estimates the T1 of water ($T1_{H2O}$) and T1 of fat ($T1_{fat}$) pools. Additionally, the confounding effects of static magnetic field inhomogeneity (Δf) and flip angle efficacy (B1) are quantified, resulting in a total of $M = 5$ MR parameters (FF, $T1_{H2O}$, $T1_{fat}$, Δf, and B1). Fingerprints are simulated using the Bloch equations with varying MRF sequence schedules of flip angles, echo times and repetition times, resulting in fingerprints of length $T = 175$.

Two dictionaries were simulated, one for training and the other for validation and testing. The training dictionary was simulated with (start:increment:stop) (0.0:0.1:1.0) for FF, (500:100:1700, 1900:200:3100) ms for $T1_{H2O}$, (200:25:400) ms for $T1_{fat}$, (−120:10:120) Hz for Δf, and (0.3:0.1:1.0) for B1. The other dictionary was simulated with (0.05:0.1:0.95) for FF, (550:200:1750, 2150:400:2950) ms for $T1_{H2O}$, (215:50:365) ms for $T1_{fat}$, (−115:20:105) Hz for Δf, and (0.35:0.1:0.95) for B1, of which randomly 20% of the entries were used for validation and the remaining 80% for testing. In total, 396000 entries were used for training, 6720 for validation, and 26880 unseen entries for testing.

2.3   Baselines and Training

We compared the INN to five baselines, one ablation and four competing NN-based methods. The ablation, termed INN_{bwd}, uses exactly the same architecture as INN but was only trained on the backward process to ablate the benefit of jointly learning the forward and backward process. The competing methods are: (i) a fully-connected NN by Cohen et al. [8] with two hidden layers, (ii) a NN by Hoppe et al. [14] consisting of four convolution layers followed by four fully-connected layer, (ii) a recurrent NN by Oksuz et al. [19] based on gated recurrent units with 100 recurrent layers followed by a fully-connected layer, and (iv) a 1-D residual convolutional NN by Song et al. [21].

All NNs were trained using a MSE loss with an Adam optimizer [15] with the learning rate chosen from $\{0.01, 0.001, 0.0005, 0.0001\}$, and $\beta_1 = 0.9, \beta_2 = 0.999$. We trained for 80 epochs and chose the batch size from $\{50, 200\}$. At each epoch, the coefficient of determination (R^2) between $\mathbf{x}$ and $\hat{\mathbf{x}}$ on the validation set was calculated and the best model was used for testing. As input, the real and imaginary parts of the complex-valued fingerprints $\mathbf{y}$ were concatenated, as commonly done [3–5,10,14], resulting in an input dimension of $2T = 350$ in all experiments. The output dimension was $M = 5$, resulting in a zero padding of $\mathbf{x}$ for the INN of $2T - M = 345$. As data augmentation, the fingerprints $\mathbf{y}$ were perturbed with random noise $\mathcal{N}(0, N^2)$. The noise standard deviation N was set to imitate signal-to-noise ratio (SNR) conditions of MRF T1-FF scans. The SNR (in dB) was defined as $20 \log_{10}(S/N)$, where S is the mean intensity of the magnitude of the magnetization at thermal equilibrium in healthy skeletal muscle. N was set to 0.003 for training, and $\mathbf{y}$ was perturbed for both the forward and backward process when training the INN. As no public code was available for the competing NNs, we implemented them in PyTorch 1.3 along with the INN. We release the code at http://www.github.com/fabianbalsiger/mrf-reconstruction-mlmir2020.

3   Experiments and Results

3.1   Backward Process: MR Parameter Estimation

The results of the MR parameter estimation from unperturbed fingerprints $\mathbf{y}$ are summarized in Table 1. The mean absolute error (MAE), the mean relative error (MRE), and the R^2 between the reference $\mathbf{x}$ and predicted $\hat{\mathbf{x}}$ MR parameters were calculated. The INN estimated all MR parameters with the highest accuracy except for the MR parameter Δf, where the INN_{bwd} yielded the best estimations in terms of MAE. Overall, all methods performed in a similar range for FF, Δf, and B1. However, a benefit in learning the Bloch simulations accentuated especially for $T1_{H2O}$ and $T1_{fat}$, where the INN outperformed all competing methods including the ablation by a considerable margin. We analyze this behaviour in more detail in Sect. 3.2.

Table 1. Mean absolute error (MAE), mean relative error (MRE), and the coefficient of determination (R^2) of the MR parameter estimation from unperturbed fingerprints. a.u.: arbitrary unit.

Metric	MR parameter	Method					
		INN	INN_{bwd}	Cohen et al.	Hoppe et al.	Oksuz et al.	Song et al.
MAE	FF	**0.008** ± 0.007	0.013 ± 0.010	0.013 ± 0.011	0.016 ± 0.012	0.015 ± 0.012	0.015 ± 0.012
	$T1_{H2O}$ (ms)	**88.9** ± 170.2	143.2 ± 249.3	140.6 ± 234.8	162.2 ± 241.8	176.0 ± 239.8	160.1 ± 243.5
	$T1_{fat}$ (ms)	**20.8** ± 19.4	27.8 ± 21.6	27.9 ± 22.0	29.0 ± 22.1	31.7 ± 23.0	28.1 ± 22.8
	Δf (Hz)	0.736 ± 0.666	**0.665** ± 0.490	0.833 ± 0.612	2.635 ± 1.503	1.380 ± 1.083	1.532 ± 1.169
	B1 (a.u.)	**0.012** ± 0.010	0.013 ± 0.010	0.015 ± 0.013	0.016 ± 0.014	0.027 ± 0.021	0.019 ± 0.014
MRE	FF (%)	**2.89** ± 4.69	4.23 ± 5.62	4.09 ± 5.02	5.64 ± 7.62	5.10 ± 6.99	6.33 ± 11.94
	$T1_{H2O}$ (%)	**6.75** ± 15.46	11.55 ± 27.23	11.47 ± 25.21	12.66 ± 25.22	13.32 ± 23.07	13.36 ± 27.42
	$T1_{fat}$ (%)	**7.48** ± 7.28	10.34 ± 9.22	10.33 ± 9.40	10.97 ± 9.80	11.96 ± 10.27	10.41 ± 9.81
	Δf (%)	**2.50** ± 5.00	2.86 ± 6.07	3.16 ± 5.91	7.52 ± 11.40	3.71 ± 5.40	5.19 ± 9.00
	B1 (%)	**1.98** ± 1.95	2.17 ± 1.88	2.56 ± 2.34	2.83 ± 2.91	4.22 ± 3.01	3.18 ± 2.69
R^2	FF	**0.999**	0.997	0.996	0.995	0.995	0.995
	$T1_{H2O}$	**0.934**	0.852	0.866	0.848	0.841	0.848
	$T1_{fat}$	**0.741**	0.604	0.596	0.574	0.508	0.582
	Δf	**1.000**	1.000	1.000	0.998	0.999	0.999
	B1	**0.994**	0.993	0.990	0.988	0.972	0.986

Robustness to noise is of considerable importance for MRF reconstruction applied to *in vivo* MRF acquisitions due to high undersampling. To simulate undersampling conditions, the performance of the INN, the INN_{bwd}, and the best competing method (Cohen et al. [8]) were analyzed under varying SNR levels, see Fig. 2. For each SNR level, we performed Monte Carlo simulations perturbing the fingerprints **y** with 100 random noise samples. It is notable that the INN more accurately and precisely estimated the MR parameters at higher SNR levels (>25 dB) than the other methods. At lower SNR levels, the differences between the methods became negligible, indicating that the benefit of learning the forward pass vanishes as the noise level increases. The plots for the MR parameters Δf and B1 look similar, and are omitted due to space constraints.

The inference time of the INN was approximately 50 ms for 1000 fingerprints, which is in-line with the competing methods. Only the training time was approximately doubled with 5 min for one epoch compared to the competing methods. The number of parameters were 0.36 million for the INN and INN_{bwd}, 0.20 million for Cohen et al. [8], 6.56 million for Hoppe et al. [14], 0.12 million for Oksuz et al. [19], and 1.49 million for Song et al. [21].

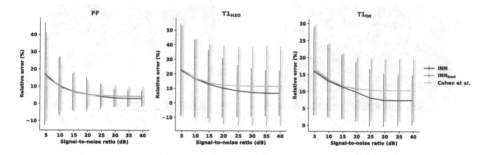

Fig. 2. Reconstruction performance in mean relative error of the INN, INN_{bwd}, and Cohen et al. [8] under varying SNR conditions for the MR parameters FF, $T1_{H2O}$, and $T1_{fat}$. Error bars indicate $\pm$ standard deviation. An SNR level of approximately 20 dB can be considered similar to an *in vivo* MRF T1-FF scan.

3.2 Forward Process: Benefit of Learning Bloch Simulations

Jointly learning the forward process mainly benefits estimating $T1_{H2O}$ and $T1_{fat}$ (cf. Table 1). To analyze this benefit, we need to introduce MR physics in presence of fat. The used sequence MRF T1-FF is designed for T1 quantification in fatty infiltrated tissues where the fat infiltration occurs at varying fractions, from no fat (FF = 0.0), to being solely fat (FF = 1.0). Unfortunately, fat infiltration, and therefore FF, greatly affects T1 quantification [18]. At FF = 1.0, $T1_{H2O}$ is not measurable as no water is present. Similarly, at FF = 0.0, $T1_{fat}$ is not measurable as no fat is present. Generally, estimating $T1_{H2O}$ is difficult at high FF values as the pooled (or global) T1 is heavily biased by the $T1_{fat}$. Contrarily, at low FF values, estimating $T1_{fat}$ is difficult as almost no fat is present. Learning the forward process could especially benefit such cases, i.e., when the information in the fingerprints is ambiguous due to MR physical restrictions. To test this assumption, we calculated the difference between the relative errors of the INN_{bwd} and INN. The heat maps in Fig. 3 show the differences for estimated $T1_{H2O}$ and $T1_{fat}$ at varying FF and $T1_{H2O}$ values. On the one hand, the forward process helped at estimating short $T1_{H2O}$ (<1000 ms) at high FF more accurately than INN_{bwd}, Fig. 3 left. Short $T1_{H2O}$ values are especially difficult to differentiate from $T1_{fat}$, as these are also very short (cf. dictionary ranges in Sect. 2.2). On the other hand, the forward process benefited the estimation of $T1_{fat}$ values at lower FF (<0.5), Fig. 3 right. At the very low FF of 0.05, the benefit diminished as it seems difficult to discriminate short $T1_{fat}$ values from longer $T1_{H2O}$ values, even when the forward process was learned. A nearly identical pattern was also obtained when comparing the INN with the method of Cohen et al. [8] (not shown). These results indicate that learning the forward process helps disentangling underlying MR physical processes.

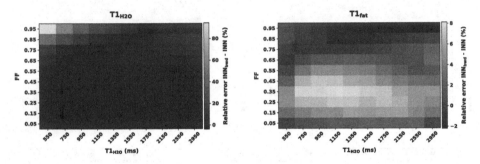

Fig. 3. Heat maps of the relative error differences between INN_{bwd} and INN for the MR parameters $T1_{H2O}$ (left) and $T1_{fat}$ (right). Positive values indicate better performance of the INN.

3.3   Relation Between the Forward and Backward Process

Learning the Bloch simulations benefits not only the MR parameter estimations but could also foster interpretability of the estimation. Due to the cyclic nature of the INN, large errors in the backward process, i.e., the error between $\mathbf{x}$ and $\hat{\mathbf{x}}$, should be associated with large errors in the forward process, i.e., the error between $\mathbf{y}$ and $\hat{\mathbf{y}}$. We tested this hypothesis by analyzing the correlation of the MRE between $\mathbf{x}$ and $\hat{\mathbf{x}}$ and the inner product between the fingerprints $\mathbf{y}$ and $\hat{\mathbf{y}}$. The association between the MRE and the inner product is shown in the scatter plot of Fig. 4. The Spearman rank-order correlation coefficient was -0.301 (p < 0.001), indicating a weak monotonic relationship. A high and a low error example are shown on the right-hand side of Fig. 4. The lower agreement between $\mathbf{y}$ and $\hat{\mathbf{y}}$ of the high error example is visually noticeable compared to the low error example. The main source of error is the $T1_{H2O}$, which is difficult to estimate at the high FF of 0.95 the fingerprint $\mathbf{y}$ was simulated with.

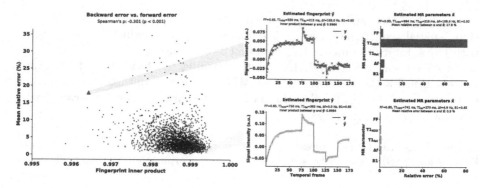

Fig. 4. Scatter plot relating the error in the forward and the backward process. For a pair $(\mathbf{x}, \mathbf{y})$, we calculated $(\hat{\mathbf{x}}, \hat{\mathbf{y}})$ using the INN and plotted the mean relative error between $\mathbf{x}$ and $\hat{\mathbf{x}}$ versus the inner product between $\mathbf{y}$ and $\hat{\mathbf{y}}$. The fingerprints and the relative errors of a high (▲) and a low (▼) error example are shown on the right-hand side. For visualization purposes, only a random subset of 10% of the data points in the scatter plot and the real part of the fingerprints were plotted. a.u.: arbitrary unit.

4   Discussion and Conclusion

We revisited NN-based MRF reconstruction by formulating it as an inverse problem. The INN allows to jointly learn the forward process from MR parameters to fingerprints and the backward process from fingerprints to MR parameters. Regarding reconstruction performance, our results suggest that learning the Bloch simulations is beneficial for MR parameter estimation.

Our experiments showed that the benefit of the INN is considerable when the information in the fingerprints is ambiguous due to MR physical restrictions. Independent of the method (invertible, fully-connected, convolutional, or recurrent) and the network size (number of parameters), FF, Δf, and B1 were nearly identically well estimated. The errors for these MR parameters were below a step size to simulate dictionaries of reasonable size for the computational intensive dictionary matching. However, this is not the case for $T1_{H2O}$ and $T1_{fat}$, where the INN performs superior. By ablation, we could attribute this performance gain to the learning of the forward process. This insight might have implications beyond T1 and FF quantification, e.g., for fast imaging with steady-state precession (FISP) sequences, where T2 relaxation time quantification is more difficult than T1 quantification [8,10,11,14]. Further, the interplay between the forward and backward process enable an enhanced interpretability of the method, which might be regarded as reconstruction uncertainty. This might be useful for MRF sequence design and optimization targeted to NN-based reconstruction.

The main limitation of this proof-of-concept study is clearly that the method was not applied to *in vivo* MRF acquisitions. Prior to doing, the behaviour of the INN under heavy noise conditions needs to be further investigated. It is currently unclear, as to why the benefit of the forward process diminishes at lower SNR levels (cf. Fig. 2). The simplest explanation is clearly the lack of enough signal, which makes MR parameter estimation difficult, independent of the method. Here, spatial regularization would most likely help [3,4], which is also possible with INNs. First attempts in this direction are promising.

In conclusion, we showed that jointly learning the forward and backward process benefits the reconstruction of MRF. INNs are suitable for such joint learning and might be a feasible alternative to the current solely backward-based NNs for MRF reconstruction.

Acknowledgement. This research was supported by the Swiss National Science Foundation (SNSF). The authors thank the NVIDIA Corporation for their GPU donation.

References

1. Adler, T.J., et al.: Uncertainty-aware performance assessment of optical imaging modalities with invertible neural networks. Int. J. Comput. Assist. Radiol. Surg. **14**(6), 997–1007 (2019). https://doi.org/10.1007/s11548-019-01939-9
2. Ardizzone, L., et al.: Analyzing inverse problems with invertible neural networks. In: International Conference on Learning Representations (2019)
3. Balsiger, F., Jungo, A., Scheidegger, O., Carlier, P.G., Reyes, M., Marty, B.: Spatially regularized parametric map reconstruction for fast magnetic resonance fingerprinting. Med. Image Anal. **64**, 101741 (2020). https://doi.org/10.1016/j.media.2020.101741
4. Balsiger, F., Scheidegger, O., Carlier, P.G., Marty, B., Reyes, M.: On the spatial and temporal influence for the reconstruction of magnetic resonance fingerprinting. In: Cardoso, M.J., et al. (eds.) Proceedings of the 2nd International Conference on Medical Imaging with Deep Learning. Proceedings of Machine Learning Research, vol. 102, pp. 27–38. PMLR, London (2019)
5. Balsiger, F., et al.: Magnetic resonance fingerprinting reconstruction via spatiotemporal convolutional neural networks. In: Knoll, F., Maier, A., Rueckert, D. (eds.) MLMIR 2018. LNCS, vol. 11074, pp. 39–46. Springer, Cham (2018). https://doi.org/10.1007/978-3-030-00129-2_5
6. Bloch, F.: Nuclear induction. Phys. Rev. **70**(7–8), 460–474 (1946). https://doi.org/10.1103/PhysRev.70.460
7. Boux, F., Forbes, F., Arbel, J., Lemasson, B., Barbier, E.: Bayesian inverse regression for vascular magnetic resonance fingerprinting. HAL preprint hal-02314026v2 (2020)
8. Cohen, O., Zhu, B., Rosen, M.S.: MR fingerprinting deep RecOnstruction NEtwork (DRONE). Magn. Reson. Med. **80**(3), 885–894 (2018). https://doi.org/10.1002/mrm.27198
9. Dinh, L., Sohl-Dickstein, J., Bengio, S.: Density estimation using real NVP. In: International Conference on Learning Representations (2017)
10. Fang, Z., et al.: Deep learning for fast and spatially-constrained tissue quantification from highly-accelerated data in magnetic resonance fingerprinting. IEEE Trans. Med. Imaging **38**(10), 2364–2374 (2019). https://doi.org/10.1109/TMI.2019.2899328

11. Fang, Z., Chen, Y., Nie, D., Lin, W., Shen, D.: RCA-U-Net: residual channel attention u-net for fast tissue quantification in magnetic resonance fingerprinting. In: Shen, D., et al. (eds.) MICCAI 2019. LNCS, vol. 11766, pp. 101–109. Springer, Cham (2019). https://doi.org/10.1007/978-3-030-32248-9_12

12. Golbabaee, M., Chen, D., Gómez, P.A., Menzel, M.I., Davies, M.E.: Geometry of deep learning for magnetic resonance fingerprinting. In: IEEE International Conference on Acoustics, Speech and Signal Processing (ICASSP), pp. 7825–7829. IEEE (2019). https://doi.org/10.1109/ICASSP.2019.8683549

13. Hoppe, E., et al.: Deep learning for magnetic resonance fingerprinting: a new approach for predicting quantitative parameter values from time series. In: Röhrig, R., Timmer, A., Binder, H., Sax, U. (eds.) German Medical Data Sciences: Visions and Bridges, vol. 243, Oldenburg, Oldenburg, pp. 202–206 (2017). https://doi.org/10.3233/978-1-61499-808-2-202

14. Hoppe, E., et al.: RinQ fingerprinting: recurrence-informed quantile networks for magnetic resonance fingerprinting. In: Shen, D., et al. (eds.) MICCAI 2019. LNCS, vol. 11766, pp. 92–100. Springer, Cham (2019). https://doi.org/10.1007/978-3-030-32248-9_11

15. Kingma, D.P., Ba, J.L.: Adam: a method for stochastic optimization. In: International Conference on Learning Representations (2015)

16. Ma, D., et al.: Magnetic resonance fingerprinting. Nature 495(7440), 187–192 (2013). https://doi.org/10.1038/nature11971

17. Marty, B., Carlier, P.G.: MR fingerprinting for water T1 and fat fraction quantification in fat infiltrated skeletal muscles. Magn. Reson. Med. 83(2), 621–634 (2019). https://doi.org/10.1002/mrm.27960

18. Marty, B., Coppa, B., Carlier, P.G.: Monitoring skeletal muscle chronic fatty degenerations with fast T1-mapping. Eur. Radiol. 28(11), 4662–4668 (2018). https://doi.org/10.1007/s00330-018-5433-z

19. Oksuz, I., et al.: Magnetic resonance fingerprinting using recurrent neural networks. In: International Symposium on Biomedical Imaging, pp. 1537–1540. IEEE (2019). https://doi.org/10.1109/ISBI.2019.8759502

20. Poorman, M.E., et al.: Magnetic resonance fingerprinting Part 1: potential uses, current challenges, and recommendations. J. Magn. Reson. Imaging 51(3), 675–692 (2019). https://doi.org/10.1002/jmri.26836

21. Song, P., Eldar, Y.C., Mazor, G., Rodrigues, M.R.: HYDRA: hybrid deep magnetic resonance fingerprinting. Med. Phys. 46(11), 4951–4969 (2019). https://doi.org/10.1002/mp.13727

Weakly-Supervised Learning for Single-Step Quantitative Susceptibility Mapping

Juan Liu[1,2(✉)] and Kevin M. Koch[1,2,3]

[1] Center for Imaging Research, Medical College of Wisconsin, Milwaukee, WI, USA
[2] Department of Biomedical Engineering, Marquette University and Medical College of Wisconsin, Milwaukee, WI, USA
juan.liu@marquette.edu
[3] Department of Radiology, Medical College of Wisconsin, Milwaukee, WI, USA

Abstract. Quantitative susceptibility mapping (QSM) utilizes MRI phase information to estimate tissue magnetic susceptibility. The generation of QSM requires solving ill-posed background field removal (BFR) and field-to-source inversion problems. Because current QSM techniques struggle to generate reliable QSM in clinical contexts, QSM clinical translation is greatly hindered. Recently, deep learning (DL) approaches for QSM reconstruction have shown impressive performance. Due to inherent non-existent ground-truth, these DL techniques use either calculation of susceptibility through multiple orientation sampling (COSMOS) maps or synthetic data for network training, which are constrained by the availability and accuracy of COSMOS maps or domain shift when training data and testing data have different domains. To address these limitations, we propose a weakly-supervised single-step QSM reconstruction method, denoted as wTFI, to directly reconstruct QSM from the total field without BFR. wTFI uses the BFR method RESHARP local fields as supervision to perform a multi-task learning of local tissue fields and QSM, and is capable of recovering magnetic susceptibility estimates near the edges of the brain where are eroded in RESHARP and realize whole brain QSM estimation. Quantitative and qualitative evaluation shows that wTFI can generate high-quality local field and susceptibility maps in a variety of neuroimaging contexts.

Keywords: QSM · Single-step QSM · Weakly-supervised learning

1 Introduction

Quantitative susceptibility mapping (QSM) can estimate tissue magnetic susceptibility values from MRI Larmor frequency sensitive phase images to provide novel image contrast [28]. To date, all QSM methods rely on a dipolar convolution that relates susceptibility sources to induced Larmor frequency offsets

© Springer Nature Switzerland AG 2020
F. Deeba et al. (Eds.): MLMIR 2020, LNCS 12450, pp. 70–81, 2020.
https://doi.org/10.1007/978-3-030-61598-7_7

[18,21], which is expressed in the k-space as bellow.

$$B(\boldsymbol{k}) = X(\boldsymbol{k}) \cdot D(\boldsymbol{k}); D(\boldsymbol{k}) = \frac{1}{3} - \frac{k_z^2}{k_x^2 + k_y^2 + k_z^2} \tag{1}$$

where $B(\boldsymbol{k})$ is the induced magnetic perturbation along the main magnetic field B_0 direction, $X(\boldsymbol{k})$ is the susceptibility distribution in the k space, $D(\boldsymbol{k})$ is the dipole kernel.

The generation of QSM requires solving two challenging ill-posed problems - (1) removal of background field contributions from sources outside the interest, (2) field-to-source inversion by solving the dipole deconvolution. Though existing BFR algorithms demonstrate excellent performance, they have several limitations, including volume erosion, inaccurate BFR close to volume boundary, and residual background leakage [23]. Incorrect BFR often introduces erroneous local field outputs and subsequently affects susceptibility quantification.

The field-to-source inversion faces technical challenges due to the singularities of the dipole kernel. Calculation of susceptibility through multiple orientation sampling (COSMOS) [15] remains the empirical gold-standard of QSM, as the additional field data sufficiently improves the conditioning of this ill-posed inversion. However, multi-orientation data acquisition is time consuming and clinically infeasible. Single-orientation QSM is preferred which is typically computed by either thresholding of the convolution operator [24,31] or sophisticated regularization methods [1,6,14,20]. In addition, several single-step QSM methods [4,17,26] have been proposed to directly estimate QSM from the total field (combined BFR and dipole inversion) to prevent potential error propagation across successive operations. However, existing QSM techniques still struggle to generate reliable QSM estimates in clinical contexts, which greatly hinders QSM clinical translation.

Recently, several deep learning (DL) QSM techniques have been proposed. For dipole inversion, QSMnet [33] and QSMGAN [5] utilized COSMOS estimates as QSM labels, while DeepQSM [2] was trained using purely synthetic data. AutoQSM [29] utilized STAR-QSM [30] estimates after SMV method [32] for BFR as QSM labels for single-step QSM learning. Though these techniques demonstrate promising performance, they have several limitations. Due to the lack of a ground-truth reference, these methods usually used COSMOS maps or synthetic data for training. However, acquiring large amounts of COSMOS data is not only expensive but also time consuming. In addition, COSMOS neglects tissue susceptibility anisotropy [11] and contains errors from BFR and image registration procedures, which compromise its value as a training label. Though synthetic data generated from the physical model provides a reliable and cost-effective way for training, the generalization needs to be addressed. In autoQSM, the robustness and accuracy of STAR-QSM could affect the performance of auto-QSM. Moreover, for DL approaches for dipole inversion only, their performance still is affected by the performance of BFR methods.

Here, we propose a weakly-supervised approach for single-step QSM, denoted as wTFI. wTFI utilizes RESHARP [27] local fields as supervision for a multi-task learning of local tissue fields and QSM. For quantitative evaluation of QSM, 9 multi-orientation datasets are utilized with comparison to TKD [24], MEDI [12], and STAR-QSM, using COSMOS result as a reference. In addition, the local tissue fields of wTFI are qualitatively compared with BFR methods, SHARP [22], RESHARP [27], PDF [13], and LBV [34] due to lack of ground-truth. More qualitative analysis is performed on single-orientation datasets and clinical datasets.

2   Method

WTFI was trained using a 3D convolutional neural network (CNN) with an encoder-decoder structure, as shown in Fig. 1. Since RESHARP produces more accurate local fields when compared with other BFR methods such as SHARP, PDF, and LBV. Local fields of RESHARP were utilized as supervision in the training paradigm. However, RESHARP suffered from brain erosion at the expense of losing information of local field and thus QSM measures in these regions. To address this problem, wTFI utilized a domain adaption technique to recover the lost information at brain edges for whole brain QSM.

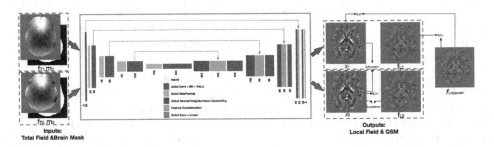

Fig. 1. Network structure of wTFI. It has an encoder-decoder structure with 9 convolutional layers (kernel size $3 \times 3 \times 3$, same padding), 9 batch normalization layers, 9 ReLU layers, 4 max pooling layers (pooling size $2 \times 2 \times 2$, strides $2 \times 2 \times 2$), 4 nearest-neighbor unsampling layers (size $2 \times 2 \times 2$), 4 feature concatenations, and 2 convolutional layer (kernel size $3 \times 3 \times 3$, linear activation).

During training, wTFI took 4 inputs (2 groups of total fields and brain masks) to get 4 outputs (2 groups of local fields and susceptibility maps). For the total field f_{T1} and brain mask m_1, the corresponding outputs were χ_1 and f_{L1}, and f_{T2} and m_2 with outputs χ_2 and f_{L2}. Define m_2 as brain mask obtained using human brain extraction tools from the MR magnitude data; m_1 as the eroded brain region after RESHARP, which has a smaller region than m_2 ($m_2 > m_1$); $f_{L_{RESHARP}}$ as RESHARP local field; f_T as the total field estimated from phase images; f_{T1} as the total field in brain region m_1, $f_{T1} = m_1 f_T$; f_{T2} as the total field in brain region m_2, $f_{T2} = m_2 f_T$.

The loss function consisted of five terms. The first loss term, L_{χ_1}, was imposed on χ_1. Leveraging the local field results from RESHARP $f_{L_{RESHARP}}$

as a weak supervision, χ_1 convoluted with the dipole kernel d should satisfy the well-established QSM inversion physical model.

$$L_{\chi_1} = \left\| m_1 W(e^{f_{L\,RESHARP}} - e^{d*\chi_1}) \right\|_2 \tag{2}$$

where W is a data-weighting factor which can be the magnitude image or noise weight matrix, $*$ is the convolution operator. Since noise is unknown and spatially variant in the local field measurements, the nonlinear dipole convolution data consistency loss was used to get more robust QSM estimates as conventional QSM methods [16,19].

Next, we included $L_{f_{L1}}$ on local field output f_{L1},

$$L_{f_{L1}} = \left\| m_1(f_{L1} - f_{L\,RESHARP}) \right\|_2 \tag{3}$$

Next, a data consistency loss was imposed on χ_2 and f_{L2}.

$$L_{consistency} = \left\| m_2 W(e^{f_{L2}} - e^{d*\chi_2}) \right\|_2 \tag{4}$$

Furthermore, define the susceptibility consistency loss between χ_1 and χ_2 inside m_1.

$$L_{\chi consistency} = \left\| m_1(\chi_1 - \chi_2) \right\|_2 \tag{5}$$

The total variation (TV) loss L_{TV} on χ_2 serves as regularization for edge preserving and denoising on the QSM.

$$L_{TV} = \left\| G_x(\chi_2) \right\|_1 + \left\| G_y(\chi_2) \right\|_1 + \left\| G_z(\chi_2) \right\|_1 \tag{6}$$

where G_x, G_y, and G_z are gradient operator in x, y, z directions.

$$L_{total} = L_{\chi_1} + \lambda_1 L_{f_{L1}} + \lambda_2 L_{consistency} + \lambda_3 L_{\chi consistency} + \lambda_4 L_{TV} \tag{7}$$

The total loss L_{total} was the weighted sum of 5 loss functions. λ_1, λ_2, λ_3, and λ_4 were loss weights.

After training, the trained DL model only took the whole brain total field f_{T2} and brain mask m_2 to get the local field and susceptibility map.

3   Experiments

Multi-orientation QSM Data. 9 QSM datasets were acquired using 5 head orientations and a 3D GRE scan with voxel size $1 \times 1 \times 1\,\mathrm{mm}^3$ from 3T MRI scanners, shared by Dr. Jongho Lee [33]. QSM data processing was implemented as below, offline GRAPPA [7] reconstruction used to reconstruct magnitude and phase images from saved k-space data, BET (FSL, FMRIB, Oxford, UK) [25] for brain mask extraction, the Laplacian method [10] for phase unwrapping, and RESHARP [27] with spherical mean radius 4 mm for BFR. COSMOS results were calculated using the 5 orientation data with image registration using FLIRT (FSL, FMRIB, Oxford, UK) [8,9]. TKD, MEDI, STAR-QSM were performed to get QSM estimates at normal head position.

For wTFI training, leave-one-out cross validation was used. For each dataset, a total of 40 scans from other 8 datasets were used for network training. wTFI was trained using patch-based neural network with patch size 96 × 96 × 96. Around 2000 patch pairs of total fields, brain masks, RESHARP local fields with patch size 96 × 96 × 96 with an overlapping of 16.6% between adjacent patches were cropped. After training, the trained DL model took the whole brain total field and brain mask of the leave-one dataset to get the local field and QSM.

QSM images at the eroded brain region were quantitatively compared with respect to COSMOS maps, using peak signal-to-noise ratio (pSNR), normalized root mean squared error (NRMSE), high-frequency error norm (HFEN), and structure similarity index (SSIM). Due to no ground-truth, the local fields of SHARP, RESHARP, PDF, LBV, and wTFI were qualitatively compared.

Single Orientation QSM Data. 200 QSM datasets were collected on a 3T MRI scanner (GE Healthcare MR750) from a susceptibility-weighted software application (SWAN, GE Healthcare). The data acquisition parameters were as follows: in-plane data matrix 320 × 256, FOV 24 cm, voxel size 0.5 × 0.5 × 2.0 mm^3, 4 TEs [10.4, 17.4, 24.4, 31.4] ms, TR 58.6 ms, and total acquisition time 4 min.

Complex multi-echo images were reconstructed from saved k-space data. The brain masks were obtained using the SPM tool [3]. After BFR using the RESHARP with spherical mean radius 4 mm, susceptibility inversion was performed using TKD, STAR-QSM, and MEDI. For wTFI training, 8000 patches of total field map, brain mask and RESHARP results from 100 datasets with patch size 128 × 128 × 64 were used for training.

Clinical Data. 150 clinical data were acquired using susceptibility-weighted angiography (SWAN, GE Healthcare, Waukesha WI) on a 3T MRI scanner (GE Healthcare MR750) with the following data acquisition parameters: in-plane data matrix 288 × 224, FOV 22 cm, slice thickness 3 mm, first TE 12.6 ms, echo spacing 4.1 ms, 7 echoes, TR 39.7 ms, pixel 244 Hz, and total acquisition time of about 2 min.

Complex multi-echo images were reconstructed from raw k-space data. The brain masks were obtained using the SPM tool [3]. RESHARP method with spherical mean radius 6 mm were used for BFR. For wTFI training, about 2000 patch pairs of total fields, brain masks and local fields from 100 datasets with patch size 128 × 128 × 64 were used.

4   Experimental Results

Multi-orientation Data. Table 1 summarized the quantitative metrics from 4 QSM methods on 9 datasets with COSMOS map as a reference. Compared to TKD, STAR-QSM, and MEDI, wTFI achieved the best metric scores in pSNR, NRMSE, and second in HFEN.

Figure 2 compared background field removal algorithms on a representative dataset. Residual background fields showed up in SHARP (b), PDF (d), and LBV

(e) results. SHARP (b) and RESHARP (c) suffered from brain erosion. PDF and LBV results showed strong shading artifacts and erroneous BFR. WTFI (f) produced RESHARP-like local fields and preserved the whole brain without erosion.

Table 1. Means and standard deviations of quantitative performance metrics from 4 reconstruction methods on 9 datasets.

	pSNR (dB)	NRMSE (%)	HFEN (%)	SSIM (0–1)
TKD	43.4 ± 0.5	91.4 ± 6.7	72.9 ± 6.6	0.831 ± 0.016
MEDI	41.5 ± 0.6	113.8 ± 7.6	100.4 ± 9.1	$\mathbf{0.902 \pm 0.016}$
STAR-QSM	45.1 ± 0.5	75.4 ± 5.4	$\mathbf{61.7 \pm 4.7}$	0.876 ± 0.016
wTFI	$\mathbf{45.3 \pm 0.5}$	$\mathbf{73.8 \pm 4.2}$	66.2 ± 3.4	0.870 ± 0.017

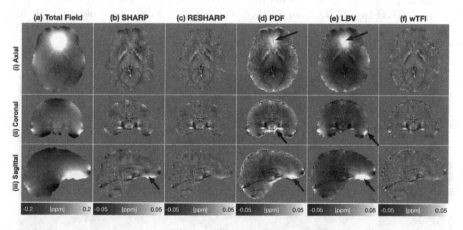

Fig. 2. Comparison of local fields of a multi-orientation dataset. SHARP (b) and RESHARP (c) suffer from brain erosion and lose local field close to brain edges. Residual background fields are visible in SHARP, PDF (d), and LBV (e) results at regions with strong background fields (black arrows). PDF and LBV show large errors in the results. WTFI (f) displays RESHARP-like local fields and preserves whole brain information.

Figure 3 displayed the QSM images. TKD (a) and MEDI (b) results showed streaking artifacts (a–b, black arrows). MEDI images lost fine details due to oversmoothing. STAR-QSM (c) images showed good quality with slight image artifacts. wTFI (d) produced high quality QSM and recovered the susceptibility information of brain edges (d, white arrows).

Single-Orientation Data. Figure 4 compared the background field removal performance of 5 methods on a single-orientation dataset. SHARP (b) and

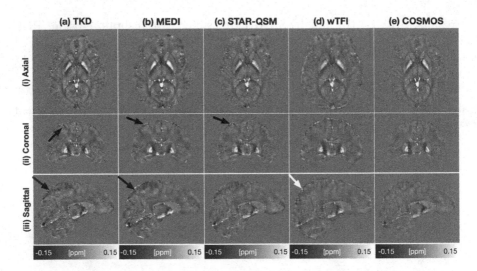

Fig. 3. Comparison of QSM of a multi-orientation dataset. Subtle streaking artifacts show up in TKD (a), MEDI, and STAR-QSM. QSM of wTFI shows clear details and invisible artifacts. Compared with other methods, wTFI is capable of recovering the susceptibility information lost in RESHARP calculation (white arrow).

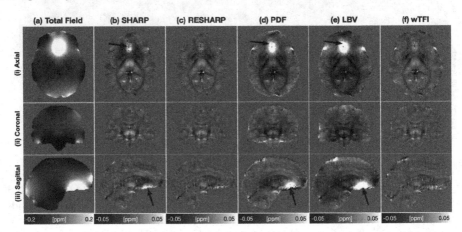

Fig. 4. Comparison of background field removal results of a single-orientation dataset. Residual background fields are observed in SHARP (b), PDF (d), LBV (e) results (b, d, e, black arrows). SHARP (b) and RESHARP (c) results have brain erosion. PDF and LBV results show strong shading artifacts in the center brain region. wTFI (f) produces RESHARP-like local fields without brain erosion.

RESHARP (c) results suffered from brain erosion. Residual background fields were observed in SHARP (b), PDF (d), LBV (e) results (b, d, e, black arrows). PDF and LBV results showed strong shading artifacts inside the brain. WTFI (f) produced RESHARP-like local fields without brain erosion.

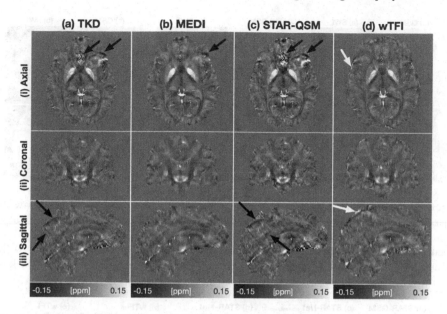

Fig. 5. Comparison of QSM estimates of a single-orientation dataset. Streaking and black shading artifacts are observed in TKD (a), MEDI (b), and STAR-QSM (c) images (black arrows). WTFI reconstructs QSM with high quality and is capable of recovering susceptibility information at brain boundaries.

Figure 5 displayed the QSM estimates. TKD (a), MEDI (b), and STAR-QSM (c) showed streaking and streaking and shading artifacts (a–c, black arrows). Due to brain erosion in RESHARP calculation, TKD, MEDI, and STAR-QSM lost susceptibility information at brain edges. Visually, wTFI (d) outperformed TKD (a), MEDI (b) and STAR-QSM (c) with invisible artifacts. In addition, WTFI was capable of recovering the susceptibility information at brain boundaries.

Clinical Data. Figure 6 displayed SWI images and susceptibility maps calculated using wTFI from 6 clinical patients. wTFI results clearly showed the brain hemorrhage, microbleeds, calcification, and vessel malformation.

Ablation Study. The ablation study was to investigate the neural network design. 3 neural networks were compared, (1) wTFI, (2) $wTFI_m$, which only used inputs f_{T1} and m_1 to get f_{L1} and χ_1. (3) STAR-Net, which used f_{T1} and m_1 as inputs, STAR-QSM maps as QSM labels for single-step QSM reconstruction and L1 loss during training.

Figure 7 showed QSM estimates of a single-orientation dataset. STAR-Net in eroded brain (b) showed STAR-QSM-like quality, but less black shadings (white arrows). Whole brain QSM of STAR-Net (c) and $wTFI_m$ (d) showed larger errors at brain boundaries when applying to whole brain f_{T2} and m_2 for QSM estimates (black arrows). This indicated the domain shift caused susceptibility quantification errors. Visual assessment showed that wTFI (e) produced better whole brain susceptibility maps.

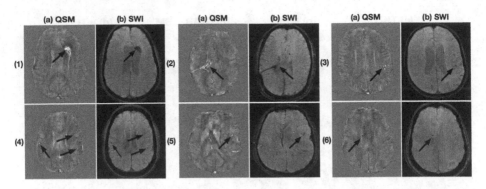

Fig. 6. QSM and SWI images from 6 clinical patients. (1) a 50-year-old patient with central nervous system (CNS) lymphoma, (2) a 30-year-old patient with encephalopathic with headache and meningitis, (3) a 80-year-old patient with Acute changes in executive functioning and metastatic bladder cancer, (4) a 32-year-old patient with metastatic non-small cell lung cancer to the brain, (5) a 48-year-old patient with left cerebral convexity meningioma, (6) a 51-year-old patient with tuberous sclerosis.

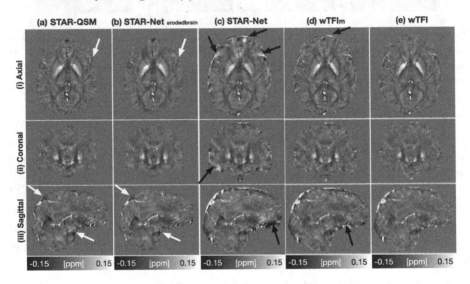

Fig. 7. Comparison of QSM of a single-orientation dataset. STAR-Net generates STAR-QSM-like images at eroded brain region m_1 (b) with less black shadings (a–b, white arrows), while at whole brain m_2 STAR-Net results (c) show sever shading artifacts (c, black arrows). $wTFI_m$ displays high quality yet subtle errors at brain boundaries, while wTFI suppresses these artifacts.

5   Discussion and Conclusion

In this work, a weakly-supervised DL method for single-step QSM was proposed. From quantitative evaluation, wTFI achieved high metric scores. However, COSMOS as reference has limitations aforementioned. Due to the lack of ground-truth for background field removal performance evaluation, qualitative assessment was performed. Based on visual assessment, wTFI outperforms conventional background field removal methods and is able to preserve the local fields at the brain boundaries. When compared with dipole inversion methods, wTFI is capable of generating high quality QSM and recovering the QSM information close to brain edges.

The proposed wTFI for QSM reconstruction approach has several advantages. First, wTFI employs RESHARP results as weak supervision and does not require QSM labels. Second, wTFI performs a multi-task learning, which estimates local fields and susceptibility map at the same time. The local field estimation can facilitate network training and be used to generate susceptibility weighted imaging. Third, wTFI is able to recover magnetic susceptibility of anatomical structures near the edges of the brain.

However, there are several limitations of wTFI. First, wTFI is still affected by the accuracy of the RESHARP results. Large spherical mean radius used in RESHARP produces better local fields but loses more information close to brain boundaries, which could introduce the difficulties of recovering the lost information. Thus, it is preferable to choose appropriate spherical mean radius to trade-off the background field removal performance and brain erosion. It might be feasible to perform multi-stage training when using RESHARP with large spherical mean radius. Second, wTFI utilizes the dipole kernel convolution physical model to infer QSM, which is constrained by the accuracy of the physical model. Future work is necessary to investigate sophisticated network architecture, evaluation of wTFI using synthetic data, and whole head single-step QSM without brain extraction.

Acknowledgement. We thank Professor Jongho Lee for sharing the multi-orientation QSM datasets.

References

1. Bilgic, B., Chatnuntawech, I., Fan, A.P., et al.: Fast image reconstruction with L2-regularization. J. Magn. Reson. Imaging **40**(1), 181–191 (2014)
2. Bollmann, S., Rasmussen, K.G.B., Kristensen, M., et al.: DeepQSM - using deep learning to solve the dipole inversion for quantitative susceptibility mapping. NeuroImage **195**, 373–383 (2019)
3. Brett, M., Anton, J.L., Valabregue, R., Poline, J.B.: Region of interest analysis using an SPM toolbox. In: 8th International Conference on Functional Mapping of the Human Brain, Sendai, p. 497 (2002)
4. Chatnuntawech, I., McDaniel, P., Cauley, S.F., et al.: Single-step quantitative susceptibility mapping with variational penalties. NMR Biomed. **30**(4), e3570 (2017)
5. Chen, Y., Jakary, A., Avadiappan, S., et al.: QSMGAN: improved quantitative susceptibility mapping using 3D generative adversarial networks with increased receptive field. NeuroImage **207**, 116389 (2019)
6. De Rochefort, L., Brown, R., Prince, M.R., et al.: Quantitative MR susceptibility mapping using piece-wise constant regularized inversion of the magnetic field. Magn. Reson. Med. **60**(4), 1003–1009 (2008)

7. Griswold, M.A., Jakob, P.M., Heidemann, R.M., et al.: Generalized autocalibrating partially parallel acquisitions (GRAPPA). Magn. Reson. Med. **47**(6), 1202–1210 (2002)
8. Jenkinson, M., Bannister, P., Brady, M., Smith, S.: Improved optimization for the robust and accurate linear registration and motion correction of brain images. Neuroimage **17**(2), 825–841 (2002)
9. Jenkinson, M., Smith, S.: A global optimisation method for robust affine registration of brain images. Med. Image Anal. **5**(2), 143–156 (2001)
10. Li, W., Wu, B., Liu, C.: Quantitative susceptibility mapping of human brain reflects spatial variation in tissue composition. NeuroImage **55**(4), 1645–1656 (2011)
11. Liu, C.: Susceptibility tensor imaging. Magn. Reson. Med. **63**(6), 1471–1477 (2010)
12. Liu, J., Liu, T., de Rochefort, L., et al.: Morphology enabled dipole inversion for quantitative susceptibility mapping using structural consistency between the magnitude image and the susceptibility map. NeuroImage **59**(3), 2560–2568 (2012)
13. Liu, T., Khalidov, I., de Rochefort, L., et al.: A novel background field removal method for MRI using projection onto dipole fields. NMR Biomed. **24**(9), 1129–1136 (2011)
14. Liu, T., Liu, J., De Rochefort, L., et al.: Morphology enabled dipole inversion (MEDI) from a single-angle acquisition: comparison with COSMOS in human brain imaging. Magn. Reson. Med. **66**(3), 777–783 (2011)
15. Liu, T., Spincemaille, P., De Rochefort, L., et al.: Calculation of susceptibility through multiple orientation sampling (COSMOS): a method for conditioning the inverse problem from measured magnetic field map to susceptibility source image in MRI. Magn. Reson. Med. **61**(1), 196–204 (2009)
16. Liu, T., Wisnieff, C., Lou, M., et al.: Nonlinear formulation of the magnetic field to source relationship for robust quantitative susceptibility mapping. Magn. Reson. Med. **69**(2), 467–476 (2013)
17. Liu, Z., Kee, Y., Zhou, D., et al.: Preconditioned total field inversion (TFI) method for quantitative susceptibility mapping. Magn. Reson. Med. **78**(1), 303–315 (2017)
18. Marques, J., Bowtell, R.: Application of a Fourier-based method for rapid calculation of field inhomogeneity due to spatial variation of magnetic susceptibility. Concepts Magn. Reson. Part B Magn. Reson. Eng. **25**(1), 65–78 (2005)
19. Polak, D., Chatnuntawech, I., Yoon, J., et al.: Nonlinear dipole inversion (NDI) enables robust quantitative susceptibility mapping (QSM). NMR Biomed., e4271 (2020)
20. de Rochefort, L., Liu, T., Kressler, B., et al.: Quantitative susceptibility map reconstruction from MR phase data using Bayesian regularization: validation and application to brain imaging. Magn. Reson. Med. **63**(1), 194–206 (2010)
21. Salomir, R., de Senneville, B.D., Moonen, C.T.: A fast calculation method for magnetic field inhomogeneity due to an arbitrary distribution of bulk susceptibility. Concepts Magn. Reson. Part B Magn. Reson. Eng. **19**(1), 26–34 (2003)
22. Schweser, F., Deistung, A., Lehr, B.W., et al.: Differentiation between diamagnetic and paramagnetic cerebral lesions based on magnetic susceptibility mapping. Med. Phys. **37**(10), 5165–5178 (2010)
23. Schweser, F., Robinson, S.D., de Rochefort, L., et al.: An illustrated comparison of processing methods for phase MRI and QSM: removal of background field contributions from sources outside the region of interest. NMR Biomed. **30**(4), e3604 (2017)
24. Shmueli, K., de Zwart, J.A., van Gelderen, P., et al.: Magnetic susceptibility mapping of brain tissue in vivo using MRI phase data. Magn. Reson. Med. **62**(6), 1510–1522 (2009)

25. Smith, S.M.: Fast robust automated brain extraction. Hum. Brain Mapp. **17**(3), 143–155 (2002)
26. Sun, H., Ma, Y., MacDonald, M.E., et al.: Whole head quantitative susceptibility mapping using a least-norm direct dipole inversion method. NeuroImage **179**, 166–175 (2018)
27. Sun, H., Wilman, A.H.: Background field removal using spherical mean value filtering and Tikhonov regularization. Magn. Reson. Med. **71**(3), 1151–1157 (2014)
28. Wang, Y., Liu, T.: Quantitative susceptibility mapping (QSM): decoding MRI data for a tissue magnetic biomarker. Magn. Reson. Med. **73**(1), 82–101 (2015)
29. Wei, H., et al.: Learning-based single-step quantitative susceptibility mapping reconstruction without brain extraction. arXiv preprint arXiv:1905.05953 (2019)
30. Wei, H., et al.: Streaking artifact reduction for quantitative susceptibility mapping of sources with large dynamic range. NMR Biomed. **28**(10), 1294–1303 (2015)
31. Wharton, S., Schäfer, A., Bowtell, R.: Susceptibility mapping in the human brain using threshold-based k-space division. Magn. Reson. Med. **63**(5), 1292–1304 (2010)
32. Wu, B., Li, W., Guidon, A., et al.: Whole brain susceptibility mapping using compressed sensing. Magn. Reson. Med. **67**(1), 137–147 (2012)
33. Yoon, J., Gong, E., Chatnuntawech, I., et al.: Quantitative susceptibility mapping using deep neural network: QSMnet. NeuroImage **179**, 199–206 (2018)
34. Zhou, D., Liu, T., Spincemaille, P., Wang, Y.: Background field removal by solving the Laplacian boundary value problem. NMR Biomed. **27**(3), 312–319 (2014)

Data-Consistency in Latent Space and Online Update Strategy to Guide GAN for Fast MRI Reconstruction

Shuo Chen[1,2], Shanhui Sun[3], Xiaoqian Huang[2], Dinggang Shen[2], Qian Wang[1], and Shu Liao[2(✉)]

[1] Institute for Medical Imaging Technology, School of Biomedical Engineering, Shanghai Jiao Tong University, Shanghai, China
wang.qian@sjtu.edu.cn
[2] Shanghai United Imaging Intelligence Co., Ltd., Shanghai, China
Dinggang.Shen@gmail.com, liaoshu.cse@gmail.com
[3] United Imaging Intelligence, Cambridge, MA 02140, USA

Abstract. Magnetic Resonance Imaging (MRI) is one of the most commonly used modalities in medical imaging, and one of the main limitations is its slow scanning and reconstruction speed. Recently, deep learning used for compressed sensing (CS) methods have been proposed to accelerate the acquisition by undersampling in the K-space and reconstruct images with neural networks. However, there are still some challenges remained: First, directly training networks based on L1/L2 distance to the target fully sampled images may lead to fuzzy reconstruction images because L1/L2 loss only enforces the overall image or patch similarity, but does not consider the local details such as anatomical sharpness. Second, Generative Adversarial Networks (GAN) can partially solve this problem. The undersampling image gets the latent space through the encoder, and the image is reconstructed by the decoder based on GAN loss, but it may generate unrealistic details by lacking of constraints in K-space domain. Third, most of the networks after training are fixed and have limited adaptation capability in the inference time, and the patient-specific information cannot be effectively used. To resolve these challenges, we proposed a new compressed sensing GAN reconstruction method, and there are two main contributions: (1) We proposed a encoder-decoder structure, which guided GAN optimization strategy data-consistency in latent space to improve the reconstruction quality such as preserving more local details and improving the anatomical sharpness while constraining GAN to follow the data distribution in the K-space to prevent the unrealistic details. (2) An online update strategy is used to find the best representation in the latent space for the underlying patient, and the reconstruct result can be further improved by incorporating the patient-specific information. Extensive experimental results show the effectiveness of our method.

Keywords: MRI reconstruction · Data-consistency · Latent space · Online update

© Springer Nature Switzerland AG 2020
F. Deeba et al. (Eds.): MLMIR 2020, LNCS 12450, pp. 82–90, 2020.
https://doi.org/10.1007/978-3-030-61598-7_8

1 Introduction

Magnetic Resonance Imaging (MRI) plays an important role in medical imaging. It has advantages such as radiation free, superior soft tissue contrast and complementary multi-sequence information compared to other image modalities. However, one of the disadvantages of MRI is its relatively slow scanning and reconstruction speed compared to other image modalities, which limits its usage in some clinical applications where image acquisition time is a critical factor.

To resolve this problem, lots of reconstruction methods were proposed. For instance, parallel imaging methods make usage of spatial sensitivity information provided by the multi-coil in order to remove the aliasing artifacts or directly reconstruct the missing information in K-space [1-4]. Also, Candes et al. [5] proposed the compressed sensing theory to overcome the limitation of the traditional Nyquist sampling theorem. It uses random projection to directly sample a small number of data points at a sampling frequency far lower than Nyquist frequency, and uses nonlinear reconstruction algorithm to recover the original signal. In 2007, Lustig et al. [6] used the random sampling of variable density, selected the traditional discrete wavelet transform for sparse transformation of image, and used the nonlinear reconstruction method of L1 norm minimization to apply compression perception to MRI field.

Recently, deep learning based image reconstruction becomes a hot research topic to use neural networks to reconstruct images. Wang and et al. [7] used convolutional neural network (CNN) to learn the mapping function between the zero-filled images in the undersampled K-space and the fully sampled images. Schlemper et al. proposed the Cascade CNN [8] structure, which cascading several CNN networks to reconstruct the images in a progressive manner. Besides, Alternating Direction Multiplier Method-Network (ADMM-Net) [9] combines ADMM algorithm and CNN for reconstruction. Most of the methods mentioned above train the network by optimizing the L1/L2 distance between the reconstructed image and the fully sampled image. However, using this strategy only will lead to fuzzy images and loss of details. To resolve this problem, The generative adversarial networks (GAN) [10] have been also adopted in training the reconstruction network. The main benefit of using GAN is it can automatically adjust the reconstruction network to provide reconstruction images with similar anatomical sharpness to the target fully sampled image through the adversarial process, and many reconstruction and image synthesis methods have been proposed based on GAN. For instance, DAGAN [11] network uses the idea of GAN in MRI reconstruction, the generator network uses the idea of U-Net [12], the discriminator network uses the network of DCGAN [13]. In the GANCS [14] network, both LSGAN [15] and CycleGAN [16] are integrated to optimize the network. Pengyue Zhang [17] proposed a method for direct reconstruction in K-space. However, the key challenge of introducing GAN in it may produce unrealistic image details, which will significantly degrade the diagnosis value of the resulting image. Another challenge in deep learning based reconstruction methods is the network, or the generative model after training is fixed when apply to the reconstruction task during the inference time, where the patient-specific information is not fully used to further improve the reconstruction quality.

Therefore, we are motivated to propose a new deep learning based compressed sensing GAN framework to resolve the above challenges. There are two main contributions in this paper: First, we propose to construct a robust and consistent low dimensional latent space representation of input images with K-space data consistency constraint to ensure that the reconstruction process from the latent space can achieve the data consistency assumption in the compressed sensing theorem. The latent space is adapted and updated with respect to each mini-batch of training images to ensure the data consistency assumption in each mini-batch of training data and use to guide the optimization of GAN in each iteration to preserve realistic details. Second, we proposed an online update mechanism to find the best representation in the encoder's latent space associated with the underlying patient, which is optimized during the inference process to improve the reconstruction quality by incorporating the patient-specific information. Different reconstruction network structures can be used in the proposed framework, and for simplicity purpose, we used U-Net as the reconstruction network in this paper. Strategies such as cascaded networks can also be used and may further improve the reconstruction quality. Extensive experiments were conducted on the publicly available dataset to compare our method to state-of-the-art reconstruction approaches, and experimental results shown that our method achieve the best reconstruction result among the compared methods.

2   Methods

There are three main stages in our proposed framework, namely the pre-training stage, the fine-training stage, and the online update stage. For simplicity purpose, U-Net is adopted as the reconstruction network to illustrate the contribution of our framework. In the following sections, details will be provided.

2.1   Pre-training Stage

In the pre-training stage, the main purpose is to construct a stable and good initialization of low dimensional latent space from the training images, and it is served as the foundation for the subsequent fine-training and online update stages. Specifically, the latent space is constructed through the encoder part of the U-Net, and it is constructed by simply optimizing the L2 distance shown in Eq. 1:

$$L_{pretrain}(\theta_E, \theta_D) = \frac{1}{2} \parallel x_t - D(E(x_{in}, \theta_E), \theta_D) \parallel_2^2, \tag{1}$$

where x_t denotes the target fully sampled image, and x_{in} denotes the input image obtained by performing the inversed Fourier transform to the downsampled K-space. $E(\cdot)$ and $D(\cdot)$ denote the encoder and decoder operation part of the U-Net, respectively. θ_E and θ_D denote the network parameters with respect to the encoder network E and the decoder network D, respectively. After training U-Net by optimizing the $L2$ loss function in Eq. 1, the low-dimensional latent space representation z of the input image x_{in} is calculated by passing x_{in} through the encoder part of U-Net, where $z = E(x_{in}, \theta_E)$.

2.2 Fine-Training Stage

In the fine-training stage, we update the weights of the encoder network E and the decoder network D of U-Net in an alternative manner. The goal to optimize E is to ensure the encoder could encode the input images to the latent space where the basic data consistency assumption in the compressed sensing theorem is fulfilled, and therefore the data consistency constraint is applied when optimizing the encoder. On the other hand, the decoder aims to restore the reconstructed images on the basis on the well constrained latent space, and the GAN together with the image similarity loss is used when optimizing the decode, while the weights of the encoder are frozen at this step to protect the latent space.

Specifically, the optimization process of the fine-training stage could be expressed by Eq. 2:

$$L_{fine}(\theta_E, \theta_D, \theta_K) = \frac{1}{2} \parallel F_u(D(E(x_{in}, \theta_E), \theta_D)) - y \parallel_2^2$$
$$+ \gamma_1 \parallel D(E(x_{in}, \theta_E), \theta_D) - x_t \parallel_2^2 \qquad (2)$$
$$+ \gamma_2 L_{GAN}(\theta_D, \theta_K),$$

where E and D denote the encoder and decoder part of U-Net, and K denote the discriminator network in GAN. In this paper, the network structure of the discriminator is the same as Pix2Pix [18]. θ_E, θ_D and θ_K denote the network parameters to be optimized with respect to networks E, D, and K, respectively. L_{GAN} denotes the GAN loss, and in this paper we used the LSGAN loss [15]. γ_1 and γ_2 are parameters controlling the weighting of the reconstruction loss and GAN loss, respectively. Where x_t is an M-dimensional data denoting the target image, and y is an N-dimensional data representing the sampled data in the K-space, where $M << N$. F_u denotes the operation of applying the Fourier transform and downsampled the data in the K-space with the same dimension as y. The weights of θ_E and θ_D are initialized based on the pre-training stage to ensure the encoder and decoder have basic representation capability in the beginning of the fine-training stage.

During the optimization process, we first optimized the encoder while freezing the weights of decoder and discriminator, and the aim of this step is to ensure the encoder fulfilling the data consistency constraint in the compressed sensing theorem, and in this case, Eq. 2 becomes:

$$L_{fine}(\theta_E) = \frac{1}{2} \parallel F_u(D(E(x_{in}, \theta_E), \theta_D)) - y \parallel_2^2 . \qquad (3)$$

It is worth pointing out that the optimization of Eq. 3 is performed on each mini-batch of data during training. Specifically, in each iteration with a new mini-batch of data, θ_E will be re-initialzed to the weights obtained right after the pre-training stage instead of using the weights obtained from the last iteration to ensure the resulting latent space is well adapted to the current mini-batch of data, which is the basic ground when optimizing the decoder, and using the weights obtained from the last iteration suffered from the risk of reaching local minima if the distribution of mini-batch data in the previous iteration is significantly different from the distribution of mini-batch data in the current iteration.

After optimizing the parameter of θ_E in the current mini-batch iteration, the latent space created by the encoder is well satisfied the data consistency constraint for the underlying data in compressed sensing theorem, and the reconstructed image will be restored from the latent space with the decoder network D. In this case, we fix θ_E and optimize θ_D and θ_K for improving the restoration capability of network D using GAN, and Eq. 2 becomes:

$$L_{fine}(\theta_D, \theta_K) = \gamma_1 \parallel D(E(x_{in}, \theta_E), \theta_D) - x_t \parallel_2^2 + \gamma_2 . L_{GAN}(\theta_D, \theta_K), \qquad (4)$$

For the decoder network D, the optimization process is different from the encoder network E. The task of D is to recover all population data from the its representation in the latent space, thus the conventional optimization strategy is used, where the weights of θ_D and θ_K are initialized based on the previous iteration. After optimizing θ_D and θ_K, a new mini-batch of data will be received, and we will re-optimize the encoder by fixing θ_D and θ_K. This optimization process continues until L_{fine} converges.

2.3   Online Update Stage

After the fine-training stage, the weights of θ_D and θ_K will be fixed during the inference of a new patient. However, the recovery capability of θ_D can be maximized only if the latent space created for the new patient fulfill the data consistency constraint following the compressed sensing theorem. In order to achieve that, the weights of the encoder θ_E need to be adaptively learnt based on the new patient input x_{new} to create the corresponding latent space, and this step can be expressed by Eq. 5:

$$L_{online}(\theta_E) = \frac{1}{2} \parallel F_u(D(E(x_{new}, \theta_E), \theta_D)) - y_{new} \parallel_2^2, \qquad (5)$$

where x_{new} denotes the M-dimensional data obtained by performing the inversed Fourier transform to the undersampled K-space data of the new patient with zero-filling, and $y_{[new}$ is the N-dimensional data representing the sampled phase encoding data in the K-space of the new patient.

After obtaining the optimum parameter θ_E^{opt} with respect to the new patient, the final reconstructed image x_{recon} could be obtained by $x_{recon} = D(E(x_{new}, \theta_E^{opt}), \theta_D)$. The online update strategy effectively incorporate the patient-specific information during the reconstruction, where the latent space of the underlying patient fulfilling the data consistency constraint by the compressed sensing theorem is adaptively created and lead to better reconstruction quality.

3   Experiments

3.1   Datasets

The data set FastMRI is adopted in our dataset, which has two image sequences: proton-density weighted image with fat suppression (PDFS) and without fat suppression (PD). FastMRI has 973 training data, 199 validation data and 108 test data. In this paper, we focus at the acceleration factor of 4, which means the K-space is downsampling a quarter of full sampling and served as input images.

3.2   Experimental Settings

In this paper, the real and imaginary part of each slice were served as input to the network. The encoder and decoder adopted in our network refers to the up and down sampling structure of U-Net [12], and the discriminator structure in Pix2Pix [18] was adopted in our GAN network. Three metrics were used to evaluate the performance of different reconstruction methods: Normalized Mean Squared Error (NMSE), Peak Signal to Noise Ratio (PSNR in dB) and the Structural Similarity Index (SSIM). The following parameter settings were used in our method: γ_1 and γ_2 controlling the weightings of the reconstruction loss and GAN loss in Eq. 2 were set to 0.5 and 0.1, respectively. The batch size during training was set to 2, and the initial learning rate was set to 0.0001, with the Adam optimizer was adopted to automatically adjust the learning rate during training. The number of iterations to optimize the parameters of encoder during the online update stage was set to 30.

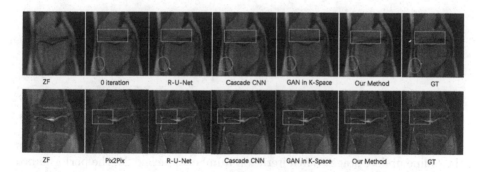

Fig. 1. Typical reconstruction results obtained by different methods: Significant differences are highlighted by red rectangles (better image contrast) and blue circles (superior local details). (Color figure online)

3.3   Experimental Results

We compared the reconstruction results obtained by different methods: Zero Filling (ZF) in the K-space, our method with 0 iteration(without update iterations), the extension of U-Net by replacing the convolution block with residual block [19] (namely the R-U-Net), Cascade CNN [8], GAN in K-space [17], and our method. Figure 1 shows typical reconstruction results obtained by different methods. It can be observed that R-U-Net lead to fuzzy reconstruction results. Our method with 0 iteration enhances the local anatomies with GAN but at the same time it brings unrealistic details. Cascade CNN produces better reconstruction results than using the GAN in K-space and R-U-Net using the hierarchical reconstruction scheme, and our method further improves the anatomical details.

The quantitative evaluation results with respect to different metrics were given in Fig. 2, respectively. It can be observed that our method consistently achieves the best result compared to other methods across all the metrics, which illustrates the effectiveness of our method. It is worth pointing out that though our method used U-Net as the reconstruction network, its reconstruction quality is significantly better than the using GAN in K-space, and it is also better than the cascaded CNN scheme [8]. Therefore, the contributions of the proposed framework are strongly reflected, and the reconstruction quality may be further improved with more advanced network structures.

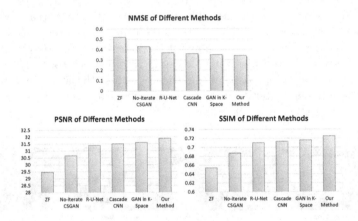

Fig. 2. Quantitative evaluation across different methods with different metrics.

Figure 3 shows the metric values calculated with respect to different iterations in the online update stage. When iteration numbers reached 20, the performance begins to converge, and normally with iteration number 30 is sufficient to produce good reconstruction results, and it only cost 124 ms in average on the Titan XP GPU, and this time cost normally satisfies the clinical workflow requirement.

In order to verify the experimental results of the three steps proposed in this paper, we conducted ablation experiments. Figure 4 show that the three steps of our method have a positive impact on MRI reconstruction tasks.

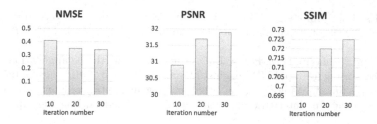

Fig. 3. Comparison of the effects of different iterations on the experimental results. Same number of iterations are used in Fine-training and Online Update Stage.

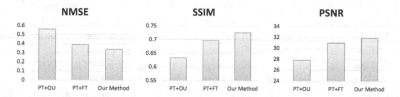

Fig. 4. We compared two ablation experiments with our methods: Pre-training (PT) Stage + Online Update (OU) Stage and Pre-training (PT) Stage + Fine-training (FT) Stage.

4   Conclusions

In this paper, we propose a new fast MRI reconstruction framework with deep learning based compressed sensing. There are two main contributions in this framework: First, the data consistent constraint in the K-space is used to guide the optimization of the encoder in each iteration to fulfill the compressed sensing theorem, and the encoder effectively builds the latent space of input data to guide the restoration process with the decoder network together with GAN. Second, an online update mechanism is proposed to incorporate the patient-specific information during the inference step of the proposed framework to further improve the reconstruction quality. Our method is compared with state-of-the-art reconstruction methods, and experimental results show that our method achieve the best reconstruction quality.

References

1. Pruessmann, K.P., Weiger, M., Scheidegger, M.B., et al.: SENSE: sensitivity encoding for fast MRI. Magn. Reson. Med. **42**, 952–962 (1999)
2. King, K.F.: ASSET-parallel imaging on the GE scanner. In: International Workshop on Parallel MRI (2004)
3. Griswold, M.A., Jakob, P.M., Heidemann, R.M., et al.: Generalized autocalibrating partially parallel acquisitions (GRAPPA). Magn. Reson. Med. **47**, 1202–1210 (2002)
4. Lustig, M., Pauly, J.: SPIRiT: iterative self-consistent parallel imaging reconstruction from arbitrary k-space. Magn. Reson. Med. **64**, 457–471 (2010)

5. Candes, E.J., Romberg, J., Tao, T., et al.: Stable signal recovery from incomplete and inaccurate measurements. Commun. Pure Appl. Math. **59**(8), 1207–1223 (2006)
6. Lustig, M., Donoho, D.L., Pauly, J.M., et al.: Sparse MRI: the application of compressed sensing for rapid MR imaging. Magn. Reson. Med. **58**(6), 1182–1195 (2007)
7. Wang, S., Su, Z., Ying, L., et al.: Accelerating magnetic resonance imaging via deep learning. In: International Symposium on Biomedical Imaging, pp. 514–517 (2016)
8. Schlemper, J., Caballero, J., Hajnal, J.V., et al.: A Deep Cascade of Convolutional Neural Networks for MR Image Reconstruction. arXiv: Computer Vision and Pattern Recognition (2017)
9. Yang, Y., Sun, J., Li, H., et al.: ADMM-CSNet: a deep learning approach for image compressive sensing. EEE Trans. Pattern Anal. Mach. Intell., 1 (2018)
10. Goodfellow, I., Pougetabadie, J., Mirza, M., et al.: Generative adversarial nets. In: Neural Information Processing Systems, pp. 2672–2680 (2014)
11. Yu, S., Dong, H., Yang, G., et al.: Deep De-Aliasing for Fast Compressive Sensing MRI. arXiv: Computer Vision and Pattern Recognition (2017)
12. Ronneberger, O., Fischer, P., Brox, T.: U-Net: convolutional networks for biomedical image segmentation. In: Navab, N., Hornegger, J., Wells, W.M., Frangi, A.F. (eds.) MICCAI 2015. LNCS, vol. 9351, pp. 234–241. Springer, Cham (2015). https://doi.org/10.1007/978-3-319-24574-4_28
13. Alec, R., Luke, M., Soumith, C.: Unsupervised Representation Learning with Deep Convolutional Generative Adversarial Networks. arXiv:1511.06434
14. Mardani, M., Gong, E., Cheng, J.Y., et al.: Deep generative adversarial networks for compressed sensing MRI. IEEE Trans. Med. Imaging **38**, 167–179 (2019)
15. Mao, X., et al.: Least squares generative adversarial networks. In: Proceedings of the IEEE International Conference on Computer Vision (2017)
16. Zhu, J.-Y., et al.: Unpaired image-to-image translation using cycle-consistent adversarial networks. In: Proceedings of the IEEE International Conference on Computer Vision (2017)
17. Zhang, P., Wang, F., Xu, W., Li, Yu.: Multi-channel generative adversarial network for parallel magnetic resonance image reconstruction in K-space. In: Frangi, A.F., Schnabel, J.A., Davatzikos, C., Alberola-López, C., Fichtinger, G. (eds.) MICCAI 2018. LNCS, vol. 11070, pp. 180–188. Springer, Cham (2018). https://doi.org/10.1007/978-3-030-00928-1_21
18. Isola, P., et al.: Image-to-image translation with conditional adversarial networks. In: Computer Vision and Pattern Recognition, pp. 5967–5976 (2017)
19. He, K., et al.: Deep residual learning for image recognition. In: Computer Vision and Pattern Recognition, pp. 770–778 (2016)

Extending LOUPE for K-Space Under-Sampling Pattern Optimization in Multi-coil MRI

Jinwei Zhang[1,3]([✉]), Hang Zhang[2,3], Alan Wang[2], Qihao Zhang[1,3],
Mert Sabuncu[1,2,3], Pascal Spincemaille[3], Thanh D. Nguyen[3],
and Yi Wang[1,3]

[1] Department of Biomedical Engineering, Cornell University, Ithaca, NY, USA
jz853@cornell.edu
[2] Department of Electrical and Computer Engineering, Cornell University, Ithaca,
NY, USA
[3] Department of Radiology, Weill Medical College of Cornell University, New York,
NY, USA

Abstract. The previously established LOUPE (Learning-based Optimization of the Under-sampling Pattern) framework for optimizing the k-space sampling pattern in MRI was extended in three folds: firstly, fully sampled multi-coil k-space data from the scanner, rather than simulated k-space data from magnitude MR images in LOUPE, was retrospectively under-sampled to optimize the under-sampling pattern of in-vivo k-space data; secondly, binary stochastic k-space sampling, rather than approximate stochastic k-space sampling of LOUPE during training, was applied together with a straight-through (ST) estimator to estimate the gradient of the threshold operation in a neural network; thirdly, modified unrolled optimization network, rather than modified U-Net in LOUPE, was used as the reconstruction network in order to reconstruct multi-coil data properly and reduce the dependency on training data. Experimental results show that when dealing with the in-vivo k-space data, unrolled optimization network with binary under-sampling block and ST estimator had better reconstruction performance compared to the ones with either U-Net reconstruction network or approximate sampling pattern optimization network, and once trained, the learned optimal sampling pattern worked better than the hand-crafted variable density sampling pattern when deployed with other conventional reconstruction methods.

Keywords: MRI · Under-sampled k-space reconstruction · Straight-through estimator · Unrolled optimization network

1 Introduction

Parallel imaging (PI) [11,22] and Compressed Sensing MRI (CS-MRI) [19] are widely used technique for acquiring and reconstructing under-sampled k-space

© Springer Nature Switzerland AG 2020
F. Deeba et al. (Eds.): MLMIR 2020, LNCS 12450, pp. 91–101, 2020.
https://doi.org/10.1007/978-3-030-61598-7_9

data thereby shortening scanning times in MRI. CS-MRI is a computational technique that suppresses incoherent noise-like artifacts introduced by random under-sampling, often via a regularized regression strategy. Combining CS-MRI with PI promises to make MRI much more accessible and affordable. Therefore, this has been an intense area of research in the past decade [9,20,28]. One major task in PI CS-MRI is designing a random under-sampling pattern, conventionally controlled by a variable-density probabilistic density function (PDF). However, the design of the 'optimal' under-sampling pattern remains an open problem for which heuristic solutions have been proposed. For example, [18] generated the sampling pattern based on the power spectrum of an existing reference dataset; [12] combined experimental design with the constrained Cramer-Rao bound to generate the context-specific sampling pattern; [10] designed a parameter-free greedy pattern selection method to find a sampling pattern that performed well on average for the MRI data in a training set.

Recently, with the success of learning based k-space reconstruction methods [1,13,24,30], a data-driven machine learning based approach called LOUPE [2] was proposed as a principled and practical solution for optimizing the under-sampling pattern in CS-MRI. In LOUPE, fully sampled k-space data was simulated from magnitude MR images and retrospective under-sampling was deployed on the simulated k-space data. A sampling pattern optimization network and a modified U-Net [23] as the under-sampled image reconstruction network were trained together in LOUPE to optimize both the k-space under-sampling pattern and reconstruction process. In the sampling pattern optimization network, one sigmoid operation was used to map the learnable weights into probability values, and a second sigmoid operation was used to approximate the non-differentiable step function for stochastic sampling, as the gradient needed to be back-propagated through such layer to update the learnable weights. After training, both optimal sampling pattern and reconstruction network were obtained. For a detailed description of LOUPE we refer the reader to [2].

In this work, we extended LOUPE in three ways. Firstly, in-house multi-coil in-vivo fully sampled T2-weighted k-space data from MR scanner was used to learn the optimal sampling pattern and reconstruction network. Secondly, modified U-Net [23] as the reconstruction network in LOUPE was extended to a modified unrolled reconstruction network with learned regularization term in order to reconstruct multi-coil data in PI with proper data consistency and reduce the dependency on training data when training cases were scarce. Thirdly, approximate stochastic sampling layer was replaced by a binary stochastic sampling layer with Straight-Through (ST) estimator [3], which was used to avoid zero gradients when back-propagating to this layer. Fully sampled data was acquired in healthy subjects. Under-sampled data was generated by retrospective under-sampling using various sampling patterns. Reconstructions were performed using different methods and compared.

2   Method

In PI CS-MRI, given an under-sampling pattern and the corresponding acquired k-space data, a reconstructed image $\hat{x}$ is obtained via minimizing the following objective function:

$$\hat{x} = \arg\min_{x} \Sigma_j^{N_c} \|UFS_j x - b_j\|_2^2 + R(x), \tag{1}$$

where x the MR image to reconstruct, S_j the coil sensitivity map of j-th coil, N_c the number of receiver coils, F the Fourier transform, U the k-space under-sampling pattern, and b_j the acquired under-sampled k-space data of the j-th coil. $R(x)$ is a regularization term, such as Total Variation (TV) [21] or wavelet [8]. The minimization in Eq. 1 is performed using iterative solvers, such as the Quasi-Newton method [7], the alternating direction method of multipliers (ADMM) [4] or the primal-dual method [5]. Equation 1 can also be mimicked by learning a parameterized mapping such as neural network from input $\{b_j\}$ to output $\hat{x}$. We denote the mapping $\{b_j\} \rightarrow \hat{x}$ using either iterative solvers or deep neural networks as $\hat{x} = \mathcal{A}(\{b_j\})$.

Our goal is to obtain an optimal under-sampling pattern $\hat{U}$ for a fixed under-sampling ratio γ from N fully sampled data through retrospective under-sampling. The mathematical formulation of this problem is:

$$\min_{U} \frac{1}{N} \Sigma_{i=1}^{N} L(x_i^*, \hat{x}_i(U)), \text{ subject to } U \in \Omega, \ \hat{x}_i(U) = \mathcal{A}(\{Ub_{ij}^*\}), \tag{2}$$

where x_i^* the i-th MR image reconstructed by direct inverse Fourier transform from fully sampled k-space data $\{b_{ij}^*\}$, $L(\cdot, \cdot)$ the loss function to measure the similarity between reconstructed image $\hat{x}_i(U)$ and fully sampled label x_i^*, Ω the constraint set of U to define how U is generated with a fixed under-sampling ratio γ. The bilevel optimization problem [6] of Eq. (2) was solved in LOUPE [2] via jointly optimizing a modified U-Net [23] as $\mathcal{A}$ and an approximate stochastic sampling process as Ω on a large volume of simulated k-space data from magnitude MR images. However, for in-vivo k-space data with multi-coil acquisition as in PI, both U-Net architecture for reconstruction and approximate stochastic sampling for pattern generation could be sub-optimal. Specifically, due to limited training size of in-vivo data and no k-space consistency imposed in U-Net, inferior reconstructions could happen in test and even training datasets. And the approximate stochastic sampling process generated fractional rather than 0–1 binary patterns during training, which might not work well during test as binary patterns should be used for realistic k-space sampling. In view of the above, we extend and improve LOUPE in terms of both reconstruction mapping $\mathcal{A}$ and sampling pattern's generating process Ω when working on in-vivo multi-coil k-space data in this work.

2.1   Unrolled Reconstruction Network

A modified residual U-Net [23] was used as the reconstruction network in LOUPE [2] to map from the zero-filled k-space reconstruction input to the fully-sampled k-space reconstruction output. U-Net works fine with simulated k-space reconstruction when enough training data of magnitude MR images are given, but as for in-vivo multi-coil k-space data, training cases are usually scarce, since fully-sampled scans are time consuming and as a result, only a few fully-sampled cases can be acquired.

To reduce the dependency on training dataset and improve the data consistency of deep learning reconstructed images, combining neural network block for the regularization term in Eq. 1 with iterative optimization scheme to solve Eq. 1 has been explored in recent years [1,13,24], which are called "unrolled optimization/reconstruction networks" in general. Prior works showed that such unrolled networks performed well for multi-coil k-space reconstruction task by means of inserting measured k-space data into the network architecture to solve Eq. 1 with a learning-based regularization. In light of the success of such unrolled reconstruction networks, we apply a modified MoDL [1] as the reconstruction network in this work. MoDL unrolled the quasi-Newton optimization scheme to solve Eq. 1 with a neural network based denoiser as the L_2 regularization term $R(x)$, and conjugate gradient (CG) descent block was applied in MoDL architecture to solve L_2 regularized problem. Besides, we will show that such unrolled network architecture also works as the skip connections for sampling pattern weights' updating as the generated pattern is connected to each intermediate CG block to perform L_2 regularized data consistency (Fig. 1).

2.2   ST Estimator for Binary Pattern

In LOUPE [2], a probabilistic pattern P_m was defined as $P_m = \frac{1}{1+e^{-a \cdot w_m}}$ with hyper-parameter a and trainable weights w_m. The binary k-space sampling pattern U was assumed to follow a Bernoulli distribution $Ber(P_m)$ independently on each k-space location. U was generated from P_m as $U = \mathbf{1}_{z<P_m}$, where $z \sim U[0,1]^{\dim(P_m)}$ and $\mathbf{1}_x$ the pointwise indicator function on the truth values of x. However, indicator function $\mathbf{1}_x$ has zero gradient almost everywhere when back-propagating through it. LOUPE addressed this issue by approximating $\mathbf{1}_{z<P_m}$ using another sigmoid function: $U \approx \frac{1}{1+e^{-b \cdot (P_m-z)}}$ with hyper-parameter b.

Although the gradient issue was solved in LOUPE, U was approximated as a fraction between $[0,1]$ on each k-space location instead of the binary pattern deployed in both test phase and realistic MR scan. As a result, binary sampling patterns generated in test phase could yield inferior performance due to such mismatch with training phase. To address this issue, binary patterns are also needed during training phase, at the same time gradient back-propagating through binary sampling layer should be properly handled. Such binary pattern generation layer can be regarded as the layer with stochastic neurons in deep learning, and several methods have been proposed to address its back-propagation [3,15]. Here we use straight through (ST) estimator [3] in the stochastic sampling layer to generate binary pattern U meanwhile addressing the zero gradient issue during back-propagation. Based on one variant of ST estimator, U is set as $\mathbf{1}_{z<P_m}$ during forward pass. When back-propagating through the stochastic

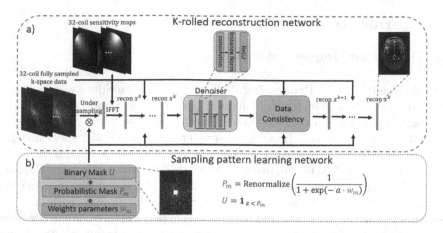

Fig. 1. Proposed network architecture consisting of a sampling pattern learning network and a K-rolled reconstruction network.

sampling layer, an ST estimator replaces the derivative factor $\frac{d\mathbf{1}_{z<P_m}}{dw_m} = 0$ with the following:

$$\frac{d\mathbf{1}_{z<P_m}}{dw_m} = \frac{dP_m}{dw_m}. \tag{3}$$

In other words, indicator function in the stochastic layer is applied at forward pass but treated as identity function during back-propagation. This ST estimator allows the network to make a yes/no decision, allowing it to picking up the top γ fraction of k-space locations most important for our task.

2.3 Network Architecture

Figure 1 shows the proposed network architecture consisting of two sub-networks: one unrolled reconstruction network and one sampling pattern learning network.

In the sampling pattern learning network (Fig. 1(b)), Renormalize($\cdot$) is a linear scaling operation to make sure the mean value of probabilistic pattern is equal to the desired under-sampling ratio γ. The binary pattern U is sampled at every forward pass in the network and once generated, it is used to retrospectively under-sample the fully sampled multi-coil k-space data.

The deep quasi-Newton network (MoDL [1]) as the unrolled reconstruction network architecture is illustrated in Fig. 1(a). In deep quasi-Newton, Denoiser + Data consistency blocks are replicated K times to mimic K quasi-Newton outer loops of solving Eq. 1 in which a neural network denoiser for $R(x)$ is applied. Five convolutional layers with skip connection [14] and instance normalization [27] are used as the denoiser and the weights are shared among blocks. The binary pattern U is used to generate zero-filled reconstruction x^0 as the input of reconstruction network and also connected to all the data consistency sub-blocks to deploy regularized optimization, which also works as the skip connection to benefit the training of pattern weights w_m.

3   Experiments

3.1   Dataset and Implementations

Data Acquisition and Processing. Fully sampled k-space data were acquired in 6 healthy subjects (5 males and 1 female; age: 30 ± 6.6 years) using a sagittal T2-weighted variable flip angle 3D fast spin echo sequence on a 3T GE scanner with a 32-channel head coil. Imaging parameters were: $256 \times 256 \times 192$ imaging matrix, $1 \, \text{mm}^3$ isotropic resolution. Coil sensitivity maps of each axial slice were calculated with ESPIRiT [25] using a $25 \times 25 \times 32$ auto-calibration k-space region. From the fully sampled data, a combined single coil image using the same coil sensitivity maps was computed to provide the ground truth label for both sampling pattern learning and reconstruction performance comparison. The central 100 slices of each subject were extracted for the training (300 slices), validation (100 slices) and test (200 slices) dataset. In addition, k-space under-sampling was performed retrospectively in the ky-kz plane for all the following experiments.

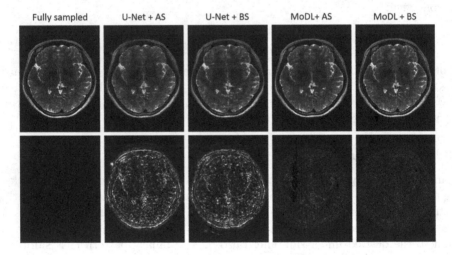

Fig. 2. Reconstruction results on one test slice by four combinations of reconstruction network and sampling pattern optimization network with 10% under-sampling ratio. First row: reconstruction results; second row: $5\times$ absolute error maps (window level: $[0, 0.5]$). MoDL + BS equipped with ST estimator had the best performance.

Training Parameters. In the sampling pattern learning network, w_m were initialized randomly, the slope factor $a = 0.25$ and the under-sampling ratio $\gamma = 10\%$. The central 25×25 k-space region remained fully sampled for each pattern. For the baseline LOUPE, a second slope factor $b = 12$ was used to approximate the binary sampling. The sampling pattern learning networks using binary sampling with ST estimator and approximated sampling were denoted as

BS (binary sampling) and AS (approximated sampling) in the following experiments. In the unrolled reconstruction network, $K = 5$ replicated blocks were applied and the denoiser was initialized randomly. For the baseline LOUPE, a residual U-Net was applied. All of the learnable parameters in Fig. 1 were trained simultaneously using the loss function: $\frac{1}{N}\Sigma_{i=1}^{N}\Sigma_{k=1}^{K}\|x_i^k - x_i^*\|_1$, where x_i^* the i-th ground truth label in the training dataset, x_i^k the k-th intermediate reconstruction ($K = 1$ in U-Net). Stochastic optimization with batch size 1 and Adam optimizer (initial learning rate: 10^{-3}) [16] was used to minimize the loss function. The number of epochs was 200. The whole training and inference procedures were implemented in PyTorch with Python version 3.7.3 on an RTX 2080Ti GPU.

Table 1. Quantitative results of Sect. 3.2

	PSNR (dB)	SSIM
U-Net+AS	32.5 ± 1.0	0.885 ± 0.016
U-Net+BS	33.0 ± 0.6	0.898 ± 0.012
MoDL+AS	41.3 ± 1.2	0.963 ± 0.015
MoDL+BS	**42.6**± 1.1	**0.968** ± 0.012

3.2 Comparison with LOUPE

Figure 2 shows the reconstruction results from one of the test subjects to demonstrate the performance improvement of the extended LOUPE over vanilla LOUPE. Four combinations of reconstruction network and sampling pattern optimization network were tested and compared. Binary sampling patterns were generated during test phase. From Fig. 2, MoDL provided better reconstruction results compared to U-Net, while for both U-Net and MoDL reconstruction networks, BS (binary sampling) gave less noisy reconstructions than AS (approximate sampling) during test phase. Quantitative comparisons in terms of PSNR (peak signal-to-noise ratio) and SSIM (structural similarity index measure [29]) are shown in Table 1, where MoDL + BS had the best performance.

Table 2. Quantitative results of Sect. 3.3

Pattern	Method	PSNR (dB)	SSIM
VD	ESPIRiT	37.5 ± 1.0	0.920 ± 0.016
	TGV	40.1 ± 0.9	0.952 ± 0.014
	MoDL	40.4 ± 0.9	0.963 ± 0.010
Learned	ESPIRiT	39.5 ± 1.1	**0.932** ± 0.018
	TGV	**42.5** ± 1.1	**0.959** ± 0.016
	MoDL	**42.6** ± 1.1	**0.968** ± 0.012

3.3   Comparison with Other Pattern

To compare the learned sampling pattern ('learned pattern' in Fig. 3, generated from MoDL + BS in Sect. 3.2) with the manually designed one with 10 % ratio, a variable density (VD) sampling pattern following a probabilistic density function

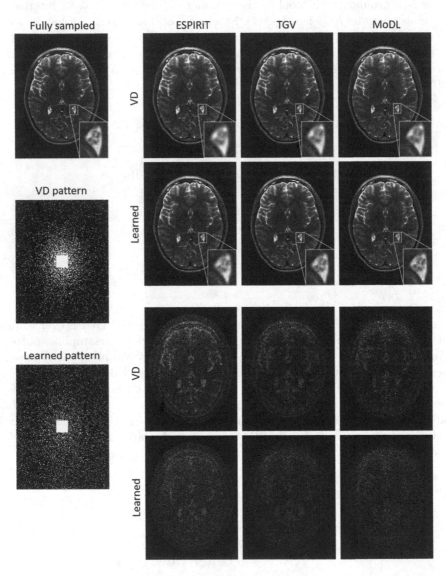

Fig. 3. Reconstruction results on another test slice using VD and learned sampling patterns with three different reconstruction methods. First two rows: reconstruction results; last two rows: corresponding 5× absolute error maps (window level:[0, 0.5]). For each reconstruction method, the learned sampling pattern produced lower global errors and sharper structural details than VD sampling pattern.

whose formula is a polynomial of the radius in k-space with tunable parameters [26] was generated ('VD pattern' in Fig. 3). ESPIRiT [25] and TGV [17] as two representative iterative methods for solving PI CS-MRI were also deployed using both sampling patterns, and the corresponding reconstruction results are shown in Fig. 3. For each reconstruction method, the learned sampling pattern captured better image depictions with lower global errors than VD pattern and the structural details as zoomed in were also sharper with the learned sampling pattern. PSNR and SSIM in Table 2 shows consistently improved performance of the learned sampling pattern over the VD pattern for each reconstruction method.

4   Conclusions

In this work, LOUPE for optimizing the k-space sampling pattern in MRI was extended by training on in-vivo multi-coil k-space data and using the unrolled network for under-sampled reconstruction and binary stochastic sampling with ST estimator for sampling pattern optimization. Experimental results show that the extended LOUPE worked better than vanilla LOUPE on in-vivo k-space data and the learned sampling pattern also performed well on other reconstruction methods. Future work includes implementing the learned sampling pattern in the pulse sequence to optimize the k-space data acquisition process prospectively.

References

1. Aggarwal, H.K., Mani, M.P., Jacob, M.: Modl: model-based deep learning architecture for inverse problems. IEEE Trans. Med. Imaging **38**(2), 394–405 (2018)
2. Bahadir, C.D., Dalca, A.V., Sabuncu, M.R.: Learning-based optimization of the under-sampling pattern in MRI. In: Chung, A.C.S., Gee, J.C., Yushkevich, P.A., Bao, S. (eds.) IPMI 2019. LNCS, vol. 11492, pp. 780–792. Springer, Cham (2019). https://doi.org/10.1007/978-3-030-20351-1_61
3. Bengio, Y., Léonard, N., Courville, A.: Estimating or propagating gradients through stochastic neurons for conditional computation. arXiv preprint arXiv:1308.3432 (2013)
4. Boyd, S., Parikh, N., Chu, E., Peleato, B., Eckstein, J., et al.: Distributed optimization and statistical learning via the alternating direction method of multipliers. Foundations Trends® Mach. Learn. **3**(1), 1–122 (2011)
5. Chambolle, A., Pock, T.: A first-order primal-dual algorithm for convex problems with applications to imaging. J. Math. Imaging Vis. **40**(1), 120–145 (2011). https://doi.org/10.1007/s10851-010-0251-1
6. Colson, B., Marcotte, P., Savard, G.: An overview of bilevel optimization. Ann. Oper. Res. **153**(1), 235–256 (2007). https://doi.org/10.1007/s10479-007-0176-2
7. Dennis Jr., J.E., Moré, J.J.: Quasi-newton methods, motivation and theory. SIAM Rev. **19**(1), 46–89 (1977)
8. Donoho, D.L., et al.: Nonlinear solution of linear inverse problems by Wavelet-Vaguelette decomposition. Appl. Comput. Harmonic Anal. **2**(2), 101–126 (1995)

9. Feng, L., et al.: Golden-angle radial sparse parallel MRI: combination of compressed sensing, parallel imaging, and golden-angle radial sampling for fast and flexible dynamic volumetric MRI. Magn. Reson. Med. **72**(3), 707–717 (2014)

10. Gözcü, B., et al.: Learning-based compressive MRI. IEEE Trans. Med. Imaging **37**(6), 1394–1406 (2018)

11. Griswold, M.A., et al.: Generalized autocalibrating partially parallel acquisitions (GRAPPA). Magn. Reson. Med. Official J. Int. Soc. Magn. Reson. Med. **47**(6), 1202–1210 (2002)

12. Haldar, J.P., Kim, D.: OEDIPUS: an experiment design framework for sparsity-constrained MRI. IEEE Trans. Med. Imaging **38**(7), 1545–1558 (2019)

13. Hammernik, K., et al.: Learning a variational network for reconstruction of accelerated MRI data. Magn. Reson. Med. **79**(6), 3055–3071 (2018)

14. He, K., Zhang, X., Ren, S., Sun, J.: Deep residual learning for image recognition. In: Proceedings of the IEEE Conference on Computer Vision and Pattern Recognition, pp. 770–778 (2016)

15. Hinton, G., Srivastava, N., Swersky, K.: Neural networks for machine learning. Coursera Video Lect. **264**(1) (2012)

16. Kingma, D.P., Ba, J.: Adam: a method for stochastic optimization. arXiv preprint arXiv:1412.6980 (2014)

17. Knoll, F., Bredies, K., Pock, T., Stollberger, R.: Second order total generalized variation (TGV) for MRI. Magn. Reson. Med. **65**(2), 480–491 (2011)

18. Knoll, F., Clason, C., Diwoky, C., Stollberger, R.: Adapted random sampling patterns for accelerated MRI. Magn. Reson. Mater. Phys. Biol. Med. **24**(1), 43–50 (2011)

19. Lustig, M., Donoho, D., Pauly, J.M.: Sparse MRI: the application of compressed sensing for rapid MR imaging. Magn. Reson. Med. Official J. Int. Soc. Magn. Reson. Med. **58**(6), 1182–1195 (2007)

20. Murphy, M., Alley, M., Demmel, J., Keutzer, K., Vasanawala, S., Lustig, M.: Fast l_1-spirit compressed sensing parallel imaging MRI: scalable parallel implementation and clinically feasible runtime. IEEE Trans. Med. Imaging **31**(6), 1250–1262 (2012)

21. Osher, S., Burger, M., Goldfarb, D., Xu, J., Yin, W.: An iterative regularization method for total variation-based image restoration. Multiscale Model. Simul. **4**(2), 460–489 (2005)

22. Pruessmann, K.P., Weiger, M., Scheidegger, M.B., Boesiger, P.: Sense: sensitivity encoding for fast MRI. Magn. Reson. Med. Official J. Int. Soc. Magn. Reson. Med. **42**(5), 952–962 (1999)

23. Ronneberger, O., Fischer, P., Brox, T.: U-Net: convolutional networks for biomedical image segmentation. In: Navab, N., Hornegger, J., Wells, W.M., Frangi, A.F. (eds.) MICCAI 2015. LNCS, vol. 9351, pp. 234–241. Springer, Cham (2015). https://doi.org/10.1007/978-3-319-24574-4_28

24. Schlemper, J., Caballero, J., Hajnal, J.V., Price, A.N., Rueckert, D.: A deep cascade of convolutional neural networks for dynamic MR image reconstruction. IEEE Trans. Med. Imaging **37**(2), 491–503 (2017)

25. Uecker, M., et al.: ESPIRiT-an eigenvalue approach to autocalibrating parallel MRI: where sense meets GRAPPA. Magn. Reson. Med. **71**(3), 990–1001 (2014)

26. Uecker, M., et al.: Berkeley advanced reconstruction toolbox. In: Proceedings of the International Society for Magnetic Resonance in Medicine, vol. 23 (2015)

27. Ulyanov, D., Vedaldi, A., Lempitsky, V.: Instance normalization: the missing ingredient for fast stylization. arXiv preprint arXiv:1607.08022 (2016)

28. Vasanawala, S., et al.: Practical parallel imaging compressed sensing MRI: summary of two years of experience in accelerating body mri of pediatric patients. In: IEEE International Symposium on Biomedical Imaging: From Nano to Macro, pp. 1039–1043. IEEE (2011)
29. Wang, Z., Bovik, A.C., Sheikh, H.R., Simoncelli, E.P.: Image quality assessment: from error visibility to structural similarity. IEEE Trans. Image Process. **13**(4), 600–612 (2004)
30. Zhang, J., et al.: Fidelity imposed network edit (fine) for solving ill-posed image reconstruction. NeuroImage **211**, 116579 (2020)

AutoSyncoder: An Adversarial AutoEncoder Framework for Multimodal MRI Synthesis

JayaChandra Raju[1]([⊠]), Balamurali Murugesan[2], Keerthi Ram[2],
and Mohanasankar Sivaprakasam[1,2]

[1] Indian Institute of Technology Madras (IITM), Chennai, India
rajubjc@gmail.com
[2] Healthcare Technology Innovation Centre (HTIC), IITM, Chennai, India

Abstract. The ability to generate multiple contrasts for the same patient is unique about MRI and of very high clinical value. In this work, we take up the problem of modality synthesis in multimodal MRI and propose an efficient, multiresolution encoder-decoder network trained like an autoencoder that can predict missed inputs at the output. This can help in avoiding the acquisition of redundant information, thereby saving time. We formulate and demonstrate our proposed AutoSyncoder network in a GAN and cyclic GAN setting, and evaluate on the BRATS-15 multimodal glioma dataset. A PSNR ranging between 29 to 30.5 dB, and SSIM over 0.88 is achieved for all the modalities, with simplistic training, thereby establishing the potential of our approach.

Keywords: Imputation · MRI · Deep learning · Image synthesis

1 Introduction

MRI is known for its ingenious and versatile imaging technique, along with the flexibility to achieve a variety of image contrasts by varying acquisition parameters. There are popular sequence types like T1-weighted, T2-weighted, proton density, etc., which accordingly produce images that show differentially highlighted tissues. When combined, these modalities are very useful to capture the conditions and underlying radiological signs.

We denote the different images obtained by applying different sequences as modalities. The possibility of synthesizing a missing modality for a subject using the other modalities of the same subject has been explored in the presented work. This could help save the time and effort put into the physical scanning of this modality. The synthesized images can also prove beneficial when the scanned image renders useless due to various factors like motion artifacts or incorrect settings while scanning.

© Springer Nature Switzerland AG 2020
F. Deeba et al. (Eds.): MLMIR 2020, LNCS 12450, pp. 102–110, 2020.
https://doi.org/10.1007/978-3-030-61598-7_10

Formulations: Multiple prior works carried out in this area indicate that information present in the acquired modalities may have relationships (or redundancy), which could be exploited to synthesize novel modalities. Methods such as modality propagation [13] and multiresolution patch regression [3] demonstrated that intensity transformations between modalities could be modeled using data-driven methods. The location-sensitive mapping approach [11] demonstrated that intensity mappings, although non-unique in general, are simpler in local volumetric neighborhoods and offer deep learning as a modeling method.

A prominent formulation explored among numerous successful deep learning approaches is **single modality input, single modality target** - applying cross-domain image translation to learn the mapping to a specified target modality, from a designated input modality [12], or from a variable input modality [2].

The next formulation of significance is **multiple modalities input, single modality target**. Such formulation entails two implications: a) parameterizing an indicator variable into the modeling, for conveying the modalities supplied [6], and b) one model corresponds to one target modality [4,14].

A singular exception is the collaborative GAN [5], which uses the specific sub-formulation that combinations of inputs being utilized to impute the missing *input*, thereby allowing **variable target modality** with a single network.

The most general formulation is to accept **variable input modalities and permit multiple targets**, as evidenced in two prior works. Multimodal modality-invariant latent representation learning [1], takes any subset of input modality combinations and synthesizes all modalities. MM-GAN [10], a single network extending the *pix2pix* framework, has N-channel input and N-channel output (statically assigned a modality each), and is trained with $2^N - 2$ combinations of input availability – zeros in input channels whenever the corresponding modalities are to be suppressed.

1.1   The AutoSyncoder

Three favorable characteristics emerge from the formulations reviewed in prior works: imputing the missing input (as in collaborative GAN), synthesizing all modalities, and avoiding indicator variable (as in multiple-target methods).

Also, we add a characteristic of avoiding training with zeros in the input (as this would impose combinatorial effort while training).

The aforementioned three considerations genuinely lead us to an N input, N output form for our solution. However, accommodating the last consideration curtails the possible combinations of inputs from $2^N - 2$ to $N - 1$

To achieve the $N - 1$ combinations without supplying zeros while training (thereby reducing training effort from $O(N)$ to $O(1)$), we propose an architecture that is simply trained like an auto-encoder. Say $N = 4$, a network of the form

$$\hat{x_a}, \hat{x_b}, \hat{x_c}, \hat{x_d} = Net(x_a, x_b, x_c, x_d) \tag{1}$$

which post-training, can provide predictions even when any one of the inputs is passed as zero, thereby achieving modality synthesis.

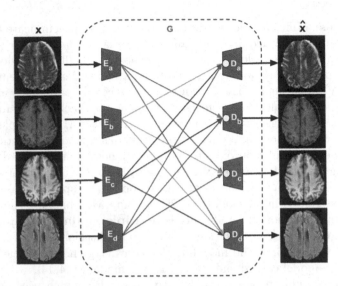

Fig. 1. Schematic of the AutoSyncoder architecture, for $N = 4$. It has encoders (E_m) and decoders (D_m) interconnected by multiresolution convolution and skip connections. Each decoder is connected to complementary encoders (both bottlenecks and skip connections). Each encoder takes in a particular modality and each decoder generates its corresponding modality. Topologically the connections in our architecture form a crown graph.

Rationale: The schematic of our proposed architecture is shown in Fig. 1. The functioning of a single stream of the AutoSyncoder can be analyzed into three modules: the encode module, meant to learn image transformation to a latent embedding, factorizing style and content from each input modality. The bottleneck module, along with inducing sparsity and avoiding trivial representations, meant to unify latent embeddings of complementary modalities. The decode module is meant to reconstruct the input modality, utilizing only the unified complementary embeddings.

Convolutional autoencoders have been used in multimodal cross-modality feature learning, such as in [7] where the modalities are image and text. In our instance, the modalities are all images, and the content is identical, with variations only in style. We do not, therefore, explicitly optimize conditions such as divergence in the embedding space. In fact we completely avoid passing information of the same modality embedding, allowing only complementary embeddings to reconstruct a modality, as discussed in the following section.

Our approach offers some additional virtues: A familiar encode-decode construction is used, with multiresolution analysis and synthesis. The architecture is **extensible** - the AutoSyncoder for $N + 1$ modalities can reuse the encode network weights from AutoSyncoder for N modalities. It is **modular**, amenable to direct use with (say) cyclic GAN, which we demonstrate (Fig. 2).

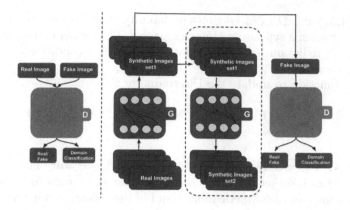

Fig. 2. An overview of our training method. On the left is the Discriminator, which has two functions: (1) D_{adv}, which will classify an image as real or fake and (2) D_{cls}, which will classify a given image to which modality it belongs to. On the right is the Generator and the Discriminator learning together to generate images that are indistinguishable from the real distribution. There is an optional flow path for the cycle, which is enclosed in a box.

2 Method

In this section we give the specifications of our basic architecture. Our network seems like an autoencoder during training. It takes all N modalities as input and produces all N modalities as output. It has a U-net-like architecture. In our case, it has N encoders and N decoders. The network has skip connections tailored specifically so that it suits the problem statement. That is, every decoder will have skip connections from its complementary encoders. Our architecture has multiple resolution levels. Each level has *conv* and *max_pooling* in case of an encoder, while in a decoder, we have *channel_stack* and *up_conv*. After the final level of encoder, we have four bottlenecks, one for each decoder. Each bottleneck will have input only from complimentary encoders, and it is passed through a convolution block. The overall Generator can be expressed as follows:

$$\hat{x}_a, \hat{x}_b, \hat{x}_c, \hat{x}_d = G(x_a, x_b, x_c, x_d) \tag{2}$$

2.1 Adversarial Training Procedure

Placing G as the Generator in an adversarial setting, the Discriminator for modality m performs two tasks: classification of $\hat{x}_m$ into real vs. synthesized, and classification of m into one of $\{a, b, c, d\}$, following the method of [2].

The total loss function to update G can be defined using adversarial loss term $1 - D_{adv}$ and negative log-likelihood term L_{cls} as

$$L_g = \sum_m \{||\hat{x}_m - x_m||_1 + (1 - ssim(\hat{x}_m, x_m)) + (1 - D_{adv}(\hat{x}_m)) + L_{cls}(\hat{x}_m)\} \tag{3}$$

where $D_{adv}(\hat{x}_m)$ and $D_{cls}(\hat{x}_m)$ denotes probability distributions produced by the Discriminator when a synthesized image is passed. The former corresponds to the probability of prediction *real*, and the latter to the class conditional probability of class m given $\hat{x}_m$.

The loss function to update the Discriminator is

$$L_d = \sum_m \{(1 - D_{adv}(x_m)) + (D_{adv}(\hat{x}_m)) + L_{cls}(x_m)\} \tag{4}$$

Cycle GAN: The second setup we show is Cyclic Generator. This setup is an extension of the above setup. In this setup, we pass the inputs to the generator and obtain a first set of synthetic images and in the following step these first set of synthetic images are passed as input to the Generator to obtain a second set of synthesized images. This can be seen as a data augmentation step in which there are two datasets (synthetic and real) that merge to one dataset towards the end of the training. Thus we have two sets of synthetic images generated by the Generator. The update process is the same as above, except that the total loss of the Generator has two extra components corresponding to the second set of synthetic images against input images. The procedure is formally stated as follows:

$$\hat{x}_a, \hat{x}_b, \hat{x}_c, \hat{x}_d = G(x_a, x_b, x_c, x_d) \tag{5}$$

$$\hat{x}_{a2}, \hat{x}_{b2}, \hat{x}_{c2}, \hat{x}_{d2} = G(\hat{x}_a, \hat{x}_b, \hat{x}_c, \hat{x}_d) \tag{6}$$

$$L_{g,cyc} = L_g + \sum_m \{||\hat{x}_{m2} - x_m||_1 + (1 - ssim(\hat{x}_{m2}, x_m))\} \tag{7}$$

The Discriminator is updated in the same way as in the previous setup.

2.2   Prediction Procedure

We took the trained Generator, and in order to synthesize a modality, we passed zeros ($\mathbf{z}$) to the corresponding encoder and passed the other three modalities for remaining encoders. The output is taken from the decoder of the corresponding modality. The procedure can be formally stated as follows:

$$\begin{aligned}
\hat{x}_a, \text{-}, \text{-}, \text{-} &= G(\mathbf{z}, x_b, x_c, x_d) \\
\text{-}, \hat{x}_b, \text{-}, \text{-} &= G(x_a, \mathbf{z}, x_c, x_d) \\
\text{-}, \text{-}, \hat{x}_c, \text{-} &= G(x_a, x_b, \mathbf{z}, x_d) \\
\text{-}, \text{-}, \text{-}, \hat{x}_d &= G(x_a, x_b, x_c, \mathbf{z})
\end{aligned} \tag{8}$$

For quantitative analysis, we used three metric functions: NMSE, PSNR, and SSIM evaluated between x_m and $\hat{x}_m$.

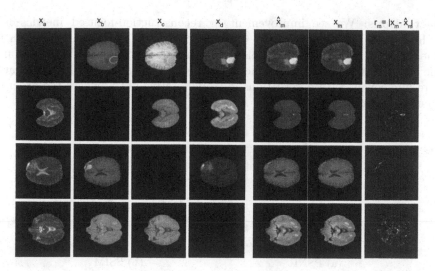

Fig. 3. Samples of test inputs (column 1 to 4), with missing inputs shown as black (zeros), and corresponding predicted $\hat{x}_m$ (column 5) in each row. In spite of no specific training with zero inputs, the predictions are close to the ground truth (x_m, in column 6). The residuals (column 7) display the pixel level differences between x_m and $\hat{x}_m$

3   Evaluation

Dataset: We have taken the BRATS 2015 dataset, [9] a multimodal MRI brain tumor dataset. It has four modalities T2, T1(Gd), T1, T2 Flair. In the dataset we conducted our experiments on HGG (High Grade Glioma) data. There are a total of 220 patient volumes in the BRATS15 training set.

Dataset Preparation: Each patient data has four modalities, and in each modality, the volume has about 155 slices. We did not use the first and last fifty slices. We have divided the 220 patient records into three sets of size 180, 20, 20 for training, validation, and test sets respectively for our experiments. Intensities are normalized linearly to be between 0 and 1 in each slice. Performance metrics are obtained on the held-out test set.

Training Details. We used the Adam optimizer with a learning rate of 0.0001, $\beta1 = 0.5$ and $\beta2 = 0.999$. The Generator is updated once for every 5 updates to the Discriminator. Using a batch size of 2 training for 100 epochs took about 5 days, occupying about 6 GB on an Nvidia GEFORCE GTX 1080 GPU.

We used gradient penalty defined similar to starGAN [2], which we extend to suit our network. The discriminator loss (Eq. 4) adding gradient penalty is:

$$L'_d = L_d - \lambda_{gp} \sum_m [(||\nabla_{\hat{x}} D_{adv}(\hat{x})|| - 1)^2] \qquad (9)$$

We use $\lambda_{gp} = 10$ in our experiments.

Observations: We see improvement in reconstruction-focused metrics like PSNR and SSIM, going from the GAN procedure to cyclic GAN, as seen in Tables 1 and 2. Visualizing residuals shows that much of detail is restored, and some modalities might be more difficult than others to synthesize. In this regard, our results concur with observations in Collaborative GAN, where the order of difficulty in synthesis is seen to be T2-Flair > T1(Gd) > T2 > T1 (Fig. 4).

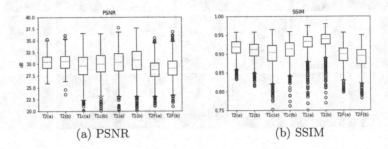

(a) PSNR   (b) SSIM

Fig. 4. PSNR and SSIM plots for various modalities using both the methods. Here a, b indicates the results obtained from GAN and Cycle-GAN respectively.

Table 1. Results on the BRATS 2015 HGG dataset for the GAN architecture. Each row shows aggregate metrics for the synthesis of the mentioned modality in test data

GAN	NMSE↓	PSNR(dB)↑	SSIM↑
T2	0.02246 ± 0.01163	30.04 ± 1.67	0.9149 ± 0.0261
T1(Gd)	0.03497 ± 0.05630	29.51 ± 2.74	0.8988 ± 0.0358
T1	0.02591 ± 0.07365	29.67 ± 3.83	0.9164 ± 0.0809
T2 Flair	0.02881 ± 0.02258	28.71 ± 2.14	0.8912 ± 0.0253

Table 2. Results on the BRATS 2015 HGG dataset for the Cycle GAN architecture.

Cycle-GAN	NMSE↓	PSNR(dB)↑	SSIM↑
T2	0.02043 ± 0.01126	30.51 ± 1.76	0.9082 ± 0.0251
T1(Gd)	0.03236 ± 0.06434	29.97 ± 2.69	0.9040 ± 0.0552
T1	0.02254 ± 0.07206	30.31 ± 3.86	0.9266 ± 0.0821
T2 Flair	0.02628 ± 0.01778	29.08 ± 2.26	0.8896 ± 0.0279

4   Discussion

Results and visuals are encouraging for a single missing input imputation, with our proposed architecture. Figure 3 shows samples from different test instances, and the ability of the network to impute the missing modality with good reconstruction metrics and minimize residuals. We see streaks of outliers in the box-whisker plots in the lower side, indicating that synthesizing some images is still challenging. Future work can focus on methods such as curriculum learning, to overcome the issue.

Limitations in the Current Study. We have shown results on the BRATS15 multimodal MRI dataset. There are other datasets used in literature, such as BRATS18, IXI, ISLES, and some prior works that have used non-registered data (within each patient recordset). Our method does not specifically impose aligned modalities as a requirement, and this prospect could be explored in future work. Another factor is that prior works also demonstrate a segmentation task, feeding the synthesized data to a standard lesion or organ segmentation network, to study performance variations pertaining to synthesis. Another related factor is the use of metrics known to correlate with radiologist perceptual opinion scores, such as VIF [8]. The construction of AutoSyncoder, as discussed in this paper permits one missing modality, a choice that offered the possibility of $O(1)$ training of the network as an autoencoder (regardless of the number of modalities N), and still be able to predict a missing input at test time. Future work can study schemes to permit multiple missing modalities in the input.

5   Conclusion

Motivated by the need to create an efficient and novel method for MRI modality synthesis, we have derived an autoencoder formulation of missing input imputation, and realized an efficient multiresolution encoder-decoder network which is simple to train. Evaluation on the BRATS15 multimodal brain MRI dataset suggests that the method is promising and opens new avenues for further research.

References

1. Chartsias, A., Joyce, T., Giuffrida, M.V., Tsaftaris, S.A.: Multimodal MR synthesis via modality-invariant latent representation. IEEE Trans. Med. Imaging **37**(3), 803–814 (2018)
2. Choi, Y., Choi, M., Kim, M., Ha, J.W., Kim, S., Choo, J.: StarGAN: Unified Generative Adversarial Networks for Multi-Domain Image-to-Image Translation, pp. 8789–8797 (2018)
3. Jog, A., Carass, A., Roy, S., Pham, D.L., Prince, J.L.: Random forest regression for magnetic resonance image synthesis. Med. Image Anal. **35**, 475–488 (2017)
4. Kim, S., Jang, H., Jang, J., Lee, Y.H., Hwang, D.: Deep-learned short tau inversion recovery imaging using multi-contrast MR images. Magn. Reson. Med. **84**, 2994–3008 (2020)
5. Lee, D., Moon, W.J., Ye, J.C.: Assessing the importance of magnetic resonance contrasts using collaborative generative adversarial networks. Nature Mach. Intell. **2**(1), 34–42 (2020)

6. Li, H., et al.: DiamondGAN: unified multi-modal generative adversarial networks for MRI sequences synthesis. In: Shen, D., et al. (eds.) MICCAI 2019. LNCS, vol. 11767, pp. 795–803. Springer, Cham (2019). https://doi.org/10.1007/978-3-030-32251-9_87

7. Liu, X., Wang, M., Zha, Z.J., Hong, R.: Cross-modality feature learning via convolutional autoencoder. ACM Trans. Multimedia Comput. Commun. Appl. **15**(1s), 7:1–7:20 (2019)

8. Mason, A., et al.: Comparison of objective image quality metrics to expert radiologists' scoring of diagnostic quality of MR images. IEEE Trans. Med. Imaging **39**(4), 1064–1072 (2020)

9. Menze, B., et al.: The multimodal brain tumor image segmentation benchmark (BRATS). IEEE Trans. Med. Imaging **34**, 1993–2024 (2014)

10. Sharma, A., Hamarneh, G.: Missing MRI pulse sequence synthesis using multimodal generative adversarial network. IEEE Trans. Med. Imaging **39**(4), 1170–1183 (2020)

11. Van Nguyen, H., Zhou, K., Vemulapalli, R.: Cross-domain synthesis of medical images using efficient location-sensitive deep network. In: Navab, N., Hornegger, J., Wells, W.M., Frangi, A.F. (eds.) MICCAI 2015. LNCS, vol. 9349, pp. 677–684. Springer, Cham (2015). https://doi.org/10.1007/978-3-319-24553-9_83

12. Yang, Q., Li, N., Zhao, Z., Fan, X., Chang, E.I.C., Xu, Y.: MRI cross-modality image-to-image translation. Sci. Rep. **10**(1), 3753 (2020)

13. Ye, D.H., Zikic, D., Glocker, B., Criminisi, A., Konukoglu, E.: Modality propagation: coherent synthesis of subject-specific scans with data-driven regularization. In: Mori, K., Sakuma, I., Sato, Y., Barillot, C., Navab, N. (eds.) MICCAI 2013. LNCS, vol. 8149, pp. 606–613. Springer, Heidelberg (2013). https://doi.org/10.1007/978-3-642-40811-3_76

14. Yurt, M., Dar, S.U.H., Erdem, A., Erdem, E., Cukur, T.: mustGAN: multi-stream generative adversarial networks for MR image synthesis. arXiv:1909.11504 [cs, eess], September 2019

Deep Learning for General Image Reconstruction

A Deep Prior Approach to Magnetic Particle Imaging

Sören Dittmer$^{(\boxtimes)}$, Tobias Kluth, Daniel Otero Baguer, and Peter Maass

Universität Bremen, Bremen, Germany
{sdittmer,tkluth,otero,pmaass}@math.uni-bremen.de

Abstract. Magnetic particle imaging (MPI) is a tracer-based imaging modality with an increasing number of potential medical applications exploiting the nonlinear magnetization behavior of magnetic nanoparticles. The image reconstruction is obtained by solving an ill-posed inverse problem requiring regularization. The number of data-driven machine learning techniques applying to inverse problems is continuously increasing. While more classical regularization techniques, e.g., variational methods, are commonly used in MPI, we focus on a novel deep image prior (DIP) approach. Initially developed for image processing tasks, it has been shown to be applicable to inverse problems. In this work, we investigate the DIP approach in the context of MPI. Its behavior is illustrated and compared to standard reconstruction methods on a 2D phantom data set obtained from the Bruker preclinical MPI system.

Keywords: Deep prior · Magnetic particle imaging · Inverse problem

1 Introduction

Magnetic particle imaging (MPI) is an imaging modality based on injecting ferromagnetic nanoparticles, which are excited by a highly dynamic applied magnetic field. The reconstruction of the resulting spatial concentration distribution of those nanoparticles is possible via exploiting the nonlinear magnetization behavior of magnetic nanoparticles [8]. More precisely, one applies a magnetic field, which is a superposition of a static gradient field, which generates a field-free-point (FFP), and a highly dynamic spatially homogeneous field, which moves the FFP in space. The mean magnetic moment of the nanoparticles in the neighborhood of the FFP generates an electro-magnetic field, which induces a measurable voltage in the so-called receive coils.

MPI benefits from a high temporal and a potentially high spatial resolution which makes it suitable for several *in-vivo* applications, such as imaging blood flow [12,31], instrument tracking [10] and guidance [27], flow estimation [7], cancer detection [33] and treatment by hyperthermia [25]. For further information on the modeling aspects, the interested reader is referred to the survey paper [17] as well as to the review article [20] for further details on the MPI methodology.

S. Dittmer and T. Kluth—Equal contribution.

© Springer Nature Switzerland AG 2020
F. Deeba et al. (Eds.): MLMIR 2020, LNCS 12450, pp. 113–122, 2020.
https://doi.org/10.1007/978-3-030-61598-7_11

Due to the nonmagnetic coating of the nanoparticles, which largely suppresses particle-particle interactions, MPI is usually modeled by a linear Fredholm integral equation of the first kind describing the relationship between particle concentration and the measured voltage. Proper modeling of the nanoparticles' mean magnetic moment in arbitrarily applied fields is still an unsolved problem such that a linear concentration-voltage relationship is commonly determined in a time-consuming calibration procedure. Reconstructing the concentration is an ill-posed problem that has been analyzed analytically in [6] for a one-dimensional and in [14] for a multi-dimensional setup. The theoretical investigations so far (based on the simplified equilibrium model [17]) conclude that in the MPI standard setup, which uses a superposition of trigonometric dynamic field excitations and linear gradient fields, is severely ill-posed.

The reconstruction problem is thus typically solved by applying a regularization method, e.g., in terms of a Tikhonov regularization [24,26,31]. For MPI this is preferably solved by using the algebraic reconstruction technique [21] combined with a nonnegativity constraint [31]. The iterative nature combined with an early stopping in the Kaczmarz iteration turns out to be beneficial for the quality in the reconstruction [16]. Further sophisticated regularization techniques such as fused lasso regularization, directional total variation, or other gradient-based methods have been applied to the MPI problem, e.g., [23]. A total-least-squares approach combined with standard Tikhonov regularization as well as a sparsity-promoting penalty term was used to improve reconstruction performance [18]. In this paper we focus on a novel machine learning approach called *deep image prior* (DIP) [2,9,11,29,30,30,32]. The idea of the DIP is to optimize the weights of an untrained neural network, which serves as an architectural prior [4], to produce a solution to a linear inverse problem.

The paper is structured as follows. We begin with a brief description of magnetic particle imaging and the idea of a DIP. We then move on to the experimental setup and the results on the MPI dataset including a comparison to standard methods. We close with a short discussion.

2   Methods

2.1   Magnetic Particle Imaging

MPI is inherently a 3D problem such that vector-valued functions remain 3D even if the domain is lower dimensional. Let $\Omega \subset E_d$, $d = 1, 2, 3$, be a bounded domain with a (strong) Lipschitz boundary $\partial\Omega$ in $E_d \subset \mathbb{R}^3$, which is a d-dimensional affine subspace. Further, let $T > 0$ denote the maximal data acquisition time and $I := (0, T)$ the time interval during which the measurement process takes place.

The induced voltage signals $u_\ell : I \to \mathbb{R}$, $\ell = 1, \ldots, L$, obtained at $L \in \mathbb{N}$ receive coils, is given by a superposition of a signal $u_{P,\ell}$ caused by the particles and the direct feedthrough $u_{E,\ell}$ (background signal) caused by the applied magnetic field $H : \mathbb{R}^3 \times I \to \mathbb{R}^3$. Analog filters $\{a_\ell\}_{\ell=1}^L$ are applied prior digitization resulting in the measured signal $v_\ell = (u_{P,\ell} + u_{E,\ell}) * a_\ell \approx u_{P,\ell} * a_\ell$ as the analog filters are chosen such that the direct feedthrough should be removed. This assumption is commonly made but may not be perfectly fulfilled [18, 28]. The inverse problem is thus to find the concentration $c : \Omega \rightarrow \mathbb{R}^+ \cup \{0\}$ from $\{v_\ell\}_{\ell=1}^L$:

$$v_\ell(t) = \int_\Omega c(x)s_\ell(x,t)\mathrm{d}x + v_{E,\ell} = S_\ell c(t) + v_{E,\ell}, \tag{1}$$

where $S_\ell : L^2(\Omega) \rightarrow L^2(I)$ and $s_\ell \in L^2(\Omega \times I)$ being the background-corrected system function.

The calibration procedure using single measurements of a small sample at predefined positions $\{x^{(i)}\}_{i=1,\dots,N} \in \Omega^N$ is outlined in the following: Let $\Gamma \subset \mathbb{R}^3$ be a reference volume placed at the origin. The concentration phantoms are given by $c^{(i)} = c_0 \chi_{x^{(i)}+\Gamma}$ for some reference concentration $c_0 > 0$. Typical choices for Γ are small cubes. If $\{x^{(i)} + \Gamma\}_{i=1,\dots,N}$ form a partition of the domain Ω, the background-corrected measurements $v_{P,\ell}^{(i)} = \frac{1}{c_0} S_\ell c^{(i)}$, $i = 1, \dots, N$, can then be used directly to characterize the system matrix S for L receive coil units. For given phantom measurements v_ℓ, $\ell = 1, \dots, L$, we build the measurement vector v analogously. Both are then given by

$$S = \begin{bmatrix} (\langle v_{P,1}^{(i)}, \psi_j \rangle)_{j \in J_1, i=1,\dots,N} \\ \vdots \\ (\langle v_{P,L}^{(i)}, \psi_j \rangle)_{j \in J_L, i=1,\dots,N} \end{bmatrix}, \quad v = \begin{bmatrix} (\langle v_1, \psi_j \rangle)_{j \in J_1} \\ \vdots \\ (\langle v_L, \psi_j \rangle)_{j \in J_L} \end{bmatrix}, \tag{2}$$

where $\{\psi_j\}_{j \in \mathbb{Z}}$ is the Fourier basis of time-periodic signals of $L^2(I)$, i.e., $\psi_j(t) = 1/\sqrt{T}(-1)^j e^{i2\pi jt/T}$, $j \in \mathbb{Z}$. $J_\ell \subset \mathbb{Z}$, $l = 1, \dots, L$, are restrictions to index sets for the purpose of a preprocessing prior the reconstruction. Let v_0 be the analogous measurement vector of the direct feedthrough. Then one obtains a measured signal from the L receive coils by $v^\delta = v + v_0 + \eta$ with noise vector η, $\|\eta\| \leq \delta$. We thus obtain a linear system of equations $Sc = v^\delta - v_0$. This linear equation system is multiplied with a diagonal matrix W with the reciprocal of the 2-norm of the respective row of S on the diagonal. This leaves us with the linear system

$$Ac = y^\delta \tag{3}$$

where the processed matrix $A = WS \in \mathbb{C}^{M \times N}$ and measurements $y^\delta = W(v - v_0) \in \mathbb{C}^M$, $M = \sum_{\ell=1}^L |J_\ell|$ are used.

Preprocessing: In MPI, there exist two standard preprocessing approaches, which are commonly combined via the index sets J_ℓ, $l = 1, \dots, L$: A bandpass approach, and SNR-type thresholding, see, e.g., [15].

2.2 Deep Image Prior

We will now briefly describe what a DIP is, for more details, we would like to refer to [2, 9, 11, 29, 30, 30, 32]. The core idea of the DIP approach is to encode the prior information into the architecture of a neural network, instead of using an additive penalty term. This means one minimizes

$$\min_{\Theta} \| A\phi_\Theta(z) - y^\delta \|^2, \tag{4}$$

where ϕ an untrained neural network with a for the construction task "conducive" architecture, $\Theta \in \mathbb{R}^q$ are the net's parameters and z a random, but fixed input, often $z \in \mathbb{R}^N$. After the optimization the reconstruction is given by $\phi_\Theta(z) \in \mathbb{R}^N$.

Network: We will now describe the setting and the architecture that we use to apply our deep image prior/regularization by architecture approach to magnetic particle imaging. Since it is not clear what a good prior for MPI would look like or how one would encode it, we will use the original DIP introduced by [29], specifically their U-net architecture. Our implementation is based on Tensorflow [1] and has, like the original DIP, the following specifications: Between the encoder and decoder part of the U-net, our skip connections have 4 channels. The convolutional encoder goes from the input to 32, 32, 64, and 128 channels, each with strides of 2×2 and filters of size 3×3. Then the convolutional decoder has the mirrored architecture (i.e., up-sampling and 128, 64, 32, 32 channels). The only modification to the original is that our network has a resize-nearest-neighbor layer to reach the desired output shape (since it is not a power of 2) and ends not on a Sigmoid, but a ReLU convolutional layer with filter-size 1 which is scaled by the factor 10^{-2}, to accommodate the magnitude of the reconstruction. The number of channels of this last layer is 3 (three 2-dimensional scans above one another) of a 2-dimensional phantom centered at the central slice of the three. The input of the network is given by a fixed Gaussian random input of size $3 \times 32 \times 32$. For further details on this architecture, we refer to [29].

2.3   Experimental Setup

We test the capability of the deep imaging prior approach to improve image reconstruction. This is done by using a dataset obtained from the Bruker preclinical MPI system. A 2D excitation in the x/y-direction is used with excitation frequencies of $2.5/102$ MHz (≈ 24.51 kHz) and $2.5/96$ MHz (≈ 26.04 kHz) resulting in a 2D Lissajous trajectory with a period of approximately 0.6528 ms. The drive field amplitude in x- and y-direction is 12 mT/μ_0, respectively. The gradient strength of the selection field is 2 T/m/μ_0 in the z-direction and -1 T/m/μ_0 in x- and y-direction. The time-dependent voltage signal is sampled with a rate of 2.5 MHz from $L = 3$ receive coil units. The discretization in time and the real-valued signal results in 817 available Fourier coefficients (for ψ_j, $j \in \{0, \dots, 816\}$) for each receive coil. Thus each system matrix S has at most $3 \cdot 817 = 2451$ rows. The *2D phantom dataset* provided in the Magnetic Particle Imaging Data Format Files (MDF) [19,22] is as follows:

The system matrix is obtained by using a cubic sample with an edge length of 1 mm. The calibration is done with Resovist tracer with a concentration of

0.25 mol/l. The field-of-view has a size of 29 mm × 29 mm × 3 mm, and the sample positions have a distance of 1 mm in each direction resulting in a size of 29 × 29 × 3 voxels, i.e., $N = 2523$ columns in the system matrix. System matrix entries are averaged over 200 repetitions, and empty scanner measurements are performed every 29 calibration scans used to build the background-corrected system matrix S. The index sets J_ℓ, $\ell = 1, 2, 3$, are based on an SNR-type thresholding with $\tau = 2$ (SNR values are also provided in the data) and the bandpass index set with the passband boundaries $b_1 = 80$ kHz and $b_2 = 625$ kHz resulting in $M = 211$ rows. It is ensured that the used phantoms are positioned within the calibrated FOV by moving an experimental platform in the desired region. The phantom measurements are averaged over 10,000 repetitions of the excitation sequence resulting in v^δ (represented as described in (2)). We use the three following phantoms:

- "4 mm": Two cylindrical glass capillary with an inner diameter of 0.7 mm filled with Resovist with a concentration of 0.25 mol/l are placed in the x/y-plane oriented in the y-direction. The heights of the tracer in the capillaries are 10 mm (left capillary) and 21 mm (right capillary). The distance between the capillaries in the x-direction is 4 mm. See also Table 1 for an illustration.
- "2 mm": Like the "4 mm" phantom with 2 mm distance in the x-direction between the glass capillary. See also Table 1 for an illustration.
- "one": The same capillaries from the "4 mm" phantom are used and arranged as the digit one. See also Table 2 for an illustration.

3   Results

We compare the DIP approach to reconstructions obtained by standard Tikhonov regularization on the dataset described in Sect. 2.3. The measured voltages are given as a sequence of complex numbers; hence we split up our **loss function** into the form

$$\|A\phi_\Theta(z) - y^\delta\|^2 = \|\Re(A)\phi_\Theta(z) - \Re(y^\delta)\|^2 + \|\Im(A)\phi_\Theta(z) - \Im(y^\delta)\|^2, \quad (5)$$

where $\Re$ and $\Im$ denote the real and imaginary parts respectively which are jointly minimized using Adam [13]. The nonconvex nature of the optimization may result in an obviously undesirable local minimum. In those cases, the optimization was restarted with a new random initialization of the network. For comparison with the DIP reconstructions, we also computed sparse and classical Tikhonov reconstructions via the minimization $\min_{c\geq 0} \|Ac - y^\delta\|^2 + \lambda\|c\|_p^p$, for $p = 1$ and $p = 2$. We solved the "$p = 2$"-case via the algebraic reconstruction technique (Kaczmarz) as generalized to allow for the constraint $c \geq 0$ by [3] and the "$p = 1$"-case via a proximal gradient descent.

We begin by presenting direct comparisons of the Kaczmarz, sparsity, and DIP reconstructions in Table 1. For the DIP, we always used early stopping after 1000 optimization steps as the images started to deteriorate slowly for more iterations. As one can see, the results presented in Table 1 are roughly on par with, if not even better than, the classical regularization methods. In particular, we would like to point out the reconstruction of the "2 mm" phantom for which only the DIP approach achieves a slight separation of the two different lines.

Table 1. Reconstruction results and photos of phantoms "4 mm" and "2 mm". The $\tilde{\lambda} = \|A\|_F^2 \lambda$ denote the penalty parameters for the Tikhonov reconstructions ($\| \cdot \|_F$ denotes the Frobenius norm) and ηs denote the learning rates used for Adam.

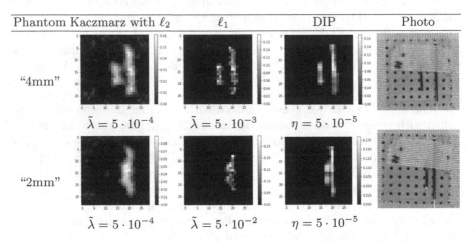

Phantom	Kaczmarz with ℓ_2	ℓ_1	DIP	Photo
"4mm"	$\tilde{\lambda} = 5 \cdot 10^{-4}$	$\tilde{\lambda} = 5 \cdot 10^{-3}$	$\eta = 5 \cdot 10^{-5}$	
"2mm"	$\tilde{\lambda} = 5 \cdot 10^{-4}$	$\tilde{\lambda} = 5 \cdot 10^{-2}$	$\eta = 5 \cdot 10^{-5}$	

We will now compare DIP reconstructions obtained with either Adam or gradient descent to a standard Landweber reconstruction of the concentration, see Table 2. To compare them we present the following quantities:

1. **Error:** The quantity we are minimizing over the iterations of the optimization, i.e., $\|Ac_k - y^\delta\|^2$ (y-axis) over k (x-axis), where k is the iteration index and c_k the reconstruction at iteration k.
2. **L-curve:** The path of the points $(\|Ac_k - y^\delta\|^2, \|c_k\|^2)$ over k, where $\|Ac_k - y^\delta\|^2$ is on the x-axis and $\|c_k\|^2$ on the y-axis. Note that these paths tend to start in the bottom right corner and end in the upper left corner.
3. **Change:** The change in c_k that happens over the course of the optimization, separately displayed for the spaces $\mathcal{N}(A)$ and $\mathcal{N}(A)^\perp$. I.e. we plot $\|P_{\mathcal{N}(A)} (c_k - c_{k-1})\|^2$ (y-axis) and $\|P_{\mathcal{N}(A)^\perp} (c_k - c_{k-1})\|^2$ (y-axis) over k (x-axis). Here P_X denotes the orthogonal projection onto the space X.
4. **Errors per Singular Value:** For each k, we plot the normalized errors associated with each of the subspaces spanned by the singular vectors, ordered by the size of their singular values, starting with the largest. I.e. for the singular value decomposition $A = U\Sigma V^*$ of A, where Σ an ordered diagonal matrix and U and V orthogonal, we plot the quantity $(V^*c_k - U^*y^\delta)_i^2$ over the indices k and i.
5. **Final Reconstruction:** c_K, where K the total number of iterations.

Table 2. Different reconstruction methods. Photo for phantom "one".

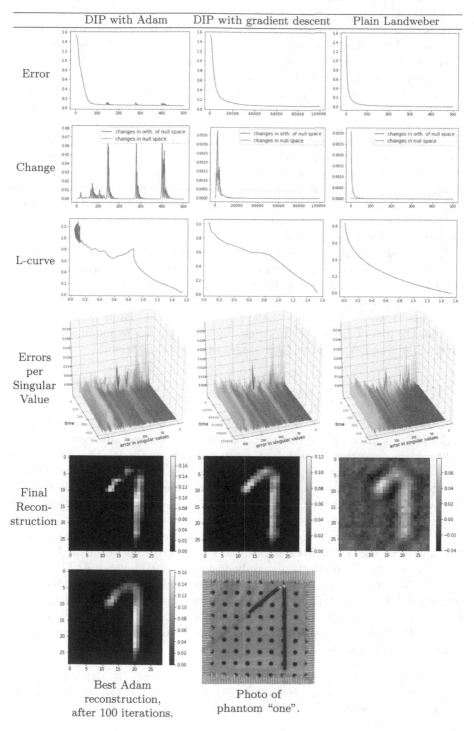

Best Adam
reconstruction,
after 100 iterations.

Photo of
phantom "one".

We want to point out that for the classical Landweber reconstruction and the Adam DIP reconstruction, we used 500 optimization steps each, but for the gradient descent DIP reconstruction, we needed 10^5 steps to obtain a reconstruction that did not change anymore. Also, the best Adam reconstruction, based on our visual judgment, was not the final one. The best one was reached after approximately 100 iterations. For both DIP reconstructions, we used a learning rate of $5 \cdot 10^{-5}$, and for Landweber one of 10^{-2}. Comparing the best reconstructions, we can see that both DIP reconstructions look quite good, although, only the gradient descent one displays the gap between the two dashes of the "1".

As one can see, the error curves of the three different methods, see Table 2, look quite similar, although the error curve of Adam has minor disruptions. Interestingly, these disruptions do lining up well with the disruptions of the "change curve" of Adam. We also found that the DIP reconstruction tends to produce good results when the choice of the optimizer and its learning rate leads to changes in the null space that are roughly on the same order of magnitude as the changes in the orthogonal of the null space.

Table 2 also presents the "errors per singular value" to illustrate the influence of so-called filter functions [5] applied to the singular values which can be used to define regularization methods. We want to point out that one can see that the DIP reconstructions allow for much bigger errors in subspaces associated with large singular values. This hints at the DIP being influential in these subspaces since, for the plain Landweber approach, one can see a flat region of small errors for the large singular values at later stages of the optimization.

4   Discussion

In this work, we successfully applied the DIP approach to MPI. The numerical results indicate that DIP reconstructions can reach comparable reconstruction quality when compared to standard reconstruction methods. The results also show that the choice of the optimization method for the training of the network and the number of iterations can have a severe influence on the reconstruction. This work is the starting point for future research in several directions. The excellent performance of iterative methods combined with an early stopping in the DIP case is similar to the behavior observed for the commonly used Kaczmarz-type method in MPI and remains to be investigated in the future.

We use a network architecture mainly designed for 2D images. As MPI is well suited for 3D tasks, future work also includes the development of a suitable architecture for 3D images. Furthermore, a comprehensive quantitative comparison of the reconstruction performance in terms of standard image quality measures (e.g., PSNR, SSIM) for various regularization methods is highly desirable for future investigations.

Acknowledgements:. The authors would like to thank P. Szwargulski and T. Knopp from the University Medical Center Hamburg-Eppendorf for their support in conducting the experiments and providing the MPI dataset. Dittmer, Kluth, and Otero

Baguer acknowledge funding by the Deutsche Forschungsgemeinschaft (DFG, German Research Foundation) - project number 281474342/GRK2224/1 "Pi^3: Parameter Identification - Analysis, Algorithms, Applications."

References

1. Abadi, M., et. al.: TensorFlow: Large-scale machine learning on heterogeneous systems (2015). https://www.tensorflow.org/, software available from tensorflow.org
2. Cheng, Z., Gadelha, M., Maji, S., Sheldon, D.: A Bayesian perspective on the deep image prior. In: The IEEE Conference on Computer Vision and Pattern Recognition (CVPR) (2019)
3. Dax, A.: On row relaxation methods for large constrained least squares problems. SIAM J. Sci. Comput. **14**(3), 570–584 (1993)
4. Dittmer, S., Kluth, T., Maass, P., Otero Baguer, D.: Regularization by architecture: a deep prior approach for inverse problems. J. Math. Imaging Vis. **62**(3), 456–470 (2019). https://doi.org/10.1007/s10851-019-00923-x
5. Engl, H.W., Hanke, M., Neubauer, A.: Regularization of Inverse Problems, Mathematics and its Applications, vol. 375. Kluwer Academic Publishers Group, Dordrecht (1996)
6. Erb, W., et al.: Mathematical analysis of the 1D model and reconstruction schemes for magnetic particle imaging. Inverse Prob. **34**(5), 21 (2018). 055012
7. Franke, J., Lacroix, R., Lehr, H., Heidenreich, M., Heinen, U., Schulz, V.: MPI flow analysis toolbox exploiting pulsed tracer information - an aneurysm phantom proof. Int. J. Magn. Part. Imaging 3(1) (2017). https://journal.iwmpi.org/index.php/iwmpi/article/view/36
8. Gleich, B., Weizenecker, J.: Tomographic imaging using the nonlinear response of magnetic particles. Nature **435**(7046), 1214–1217 (2005)
9. Gong, K., Han, P., El Fakhri, G., Ma, C., Li, Q.: Arterial spin labeling mr image denoising and reconstruction using unsupervised deep learning. NMR Biomed. e4224 (2019)
10. Haegele, J., et al.: Magnetic particle imaging: visualization of instruments for cardiovascular intervention. Radiology **265**(3), 933–938 (2012)
11. Hashimoto, F., Ohba, H., Ote, K., Teramoto, A., Tsukada, H.: Dynamic pet image denoising using deep convolutional neural networks without prior training datasets. IEEE Access **7**, 96594–96603 (2019)
12. Khandhar, A., et al.: Evaluation of peg-coated iron oxide nanoparticles as blood pool tracers for preclinical magnetic particle imaging. Nanoscale **9**(3), 1299–1306 (2017)
13. Kingma, D.P., Ba, J.: Adam: a method for stochastic optimization. arXiv preprint arXiv:1412.6980 (2014)
14. Kluth, T.: Mathematical models for magnetic particle imaging. Inverse Prob. **34**(8), 083001 (2018)
15. Kluth, T., Jin, B.: Enhanced reconstruction in magnetic particle imaging by whitening and randomized SVD approximation. Phys. Med. Biol. **64**(12), 125026 (2019)
16. Kluth, T., Jin, B.: L1 data fitting for robust numerical reconstruction in magnetic particle imaging: quantitative evaluation on Open MPI dataset. arXiv: 2001.06083 (2020, preprint)
17. Kluth, T., Jin, B., Li, G.: On the degree of ill-posedness of multi-dimensional magnetic particle imaging. Inverse Prob. **34**(9), 095006 (2018)

18. Kluth, T., Maass, P.: Model uncertainty in magnetic particle imaging: nonlinear problem formulation and model-based sparse reconstruction. Int. J. Magn. Part. Imaging **3**(2), 10 (2017). ID 1707004
19. Knopp, T.: Github MDF. https://github.com/MagneticParticleImaging/MDF. Accessed 16 Nov 2018
20. Knopp, T., Gdaniec, N., Möddel, M.: Magnetic particle imaging: from proof of principle to preclinical applications. Phys. Med. Biol. **62**(14), R124 (2017)
21. Knopp, T., Hofmann, M.: Online reconstruction of 3D magnetic particle imaging data. Phys. Med. Biol. **61**(11), N257–67 (2016)
22. Knopp, T., et al.: MDF: magnetic particle imaging data format, pp. 1–15. ArXiv e-prints 1602.06072v6, January 2018. http://arxiv.org/abs/1602.06072v6, article, MDF
23. Konkle, J., Goodwill, P., Hensley, D., Orendorff, R., Lustig, M., Conolly, S.: A convex formulation for magnetic particle imaging x-space reconstruction. PLoS One **10**(10), e0140137 (2015)
24. Lampe, J., et al.: Fast reconstruction in magnetic particle imaging. Phys. Med. Biol. **57**(4), 1113–1134 (2012)
25. Murase, K., et al.: Usefulness of magnetic particle imaging for predicting the therapeutic effect of magnetic hyperthermia. Open J. Med. Imaging **5**(02), 85 (2015)
26. Rahmer, J., Weizenecker, J., Gleich, B., Borgert, J.: Analysis of a 3-D system function measured for magnetic particle imaging. IEEE Trans. Med. Imaging **31**(6), 1289–1299 (2012)
27. Salamon, J., et al.: Magnetic particle/magnetic resonance imaging: in-vitro MPI-guided real time catheter tracking and 4D angioplasty using a road map and blood pool tracer approach. PloS One **11**(6), 14 (2016). e0156899
28. Them, K., et al.: Sensitivity enhancement in magnetic particle imaging by background subtraction. IEEE Trans. Med. Imag. **35**(3), 893–900 (2016)
29. Ulyanov, D., Vedaldi, A., Lempitsky, V.S.: Deep image prior. CoRR abs/1711.10925 (2017). http://arxiv.org/abs/1711.10925
30. Van Veen, D., Jalal, A., Price, E., Vishwanath, S., Dimakis, A.G.: Compressed sensing with deep image prior and learned regularization. arXiv preprint arXiv:1806.06438 (2018)
31. Weizenecker, J., Gleich, B., Rahmer, J., Dahnke, H., Borgert, J.: Three-dimensional real-time in vivo magnetic particle imaging. Phys. Med. Biol. **54**(5), L1 (2009)
32. Yokota, T., Kawai, K., Sakata, M., Kimura, Y., Hontani, H.: Dynamic pet image reconstruction using nonnegative matrix factorization incorporated with deep image prior. In: Proceedings of the IEEE International Conference on Computer Vision, pp. 3126–3135 (2019)
33. Yu, E.Y., et al.: Magnetic particle imaging: a novel in vivo imaging platform for cancer detection. Nano Lett. **17**(3), 1648–1654 (2017)

End-To-End Convolutional Neural Network for 3D Reconstruction of Knee Bones from Bi-planar X-Ray Images

Yoni Kasten[1,2], Daniel Doktofsky[1], and Ilya Kovler[1(✉)]

[1] RSIP Vision, Shamai 15, Jerusalem, Israel
ilya.kovler@rsipvision.com
[2] Department of Computer Science, Weizmann Institute of Science, Rehovot, Israel
https://www.rsipvision.com

Abstract. We present an end-to-end Convolutional Neural Network (CNN) approach for 3D reconstruction of knee bones directly from two bi-planar X-ray images. Clinically, capturing the 3D models of the bones is crucial for surgical planning, implant fitting, and postoperative evaluation. X-ray imaging significantly reduces the exposure of patients to ionizing radiation compared to Computer Tomography (CT) imaging, and is much more common and inexpensive compared to Magnetic Resonance Imaging (MRI) scanners. However, retrieving 3D models from such 2D scans is extremely challenging. In contrast to the common approach of statistically modeling the shape of each bone, our deep network learns the distribution of the bones' shapes directly from the training images. We train our model with both supervised and unsupervised losses using Digitally Reconstructed Radiograph (DRR) images generated from CT scans. To apply our model to X-Ray data, we use style transfer to transform between X-Ray and DRR modalities. As a result, at test time, without further optimization, our solution directly outputs a 3D reconstruction from a pair of bi-planar X-ray images, while preserving geometric constraints. Our results indicate that our deep learning model is very efficient, generalizes well and produces high quality reconstructions.

Keywords: 3D reconstruction · X-ray imaging · Deep learning · Patient specific planning

1 Introduction

3D reconstruction of knee bones is an important step for various clinical applications. It may be used for surgical planning, precise implant selection, patient

---

Y. Kasten, D. Doktofsky and I. Kovler—Equal contributors.
This work has been done at RSIP Vision.

Electronic supplementary material The online version of this chapter (https://doi.org/10.1007/978-3-030-61598-7_12) contains supplementary material, which is available to authorized users.

© Springer Nature Switzerland AG 2020
F. Deeba et al. (Eds.): MLMIR 2020, LNCS 12450, pp. 123–133, 2020.
https://doi.org/10.1007/978-3-030-61598-7_12

specific implant manufacturing or intraoperative jig printing which perfectly fits the anatomy. X-ray images are often used due to their wide availability, lower price, short scanning time and lower levels of ionizing radiation compared to CT scanners. However, since X-ray images provide only 2D information, some prior knowledge must be incorporated in order to extract the missing dimension. Previous approaches [5,7,8,17] use Statistical Shape Models (SSM) or Statistical Shape and Intensity Models (SSIM) for reconstructing bones from X-ray images. However, optimization for the deformable model parameters might be slow and needs a good initialization point to avoid local maxima [15,22].

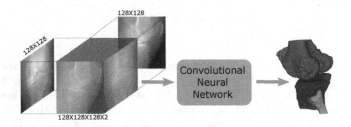

Fig. 1. General scheme: AP and lateral X-ray scans of the knee joint are replicated into a 3D array, each on a different channel (green and red for illustration). A CNN then predicts a 3D segmentation map of the bone classes which is used for 3D reconstruction of the bones. (Color figure online)

In this paper, we present a novel end-to-end deep learning approach for 3D reconstruction of knee joints from two bi-planar X-ray images. The overall scheme is presented in Fig. 1. CNNs have recently proven very effective for various types of tasks [18], including image segmentation and classification. However, implementing 3D reconstruction from two or more 2D images using a deep learning approach remains a challenging task, due to the difficulty of representing a dimensional enlargement in multi-view settings with standard differentiable layers. Moreover, due to the transparent nature of X-ray images, matching surface points between multi-views for dense reconstruction is extremely challenging compared to the standard multi-view setting [11].

We address these challenges by introducing a dimensional enlargement approach that given two bi-planar X-rays back-projects each pair of corresponding epipolar lines into a two-channeled epipolar plane. This results in a 3D volume that contains all the information observed from the two X-ray images, while preserving the two-view geometric constraints. We combine this representation with a deep learning architecture that outputs 3D models of the different bones. The experiments show the utilization of our approach for 3D reconstruction of knee joint. We strongly believe that our method paves the way for future research in deep learning based 3D modeling of bones from X-ray scans. In contrast to SSM based methods, our method does not require an initialization and runs in 0.5 s while a standard SSM optimization for one knee bone takes about 4.88 s.

2   Related Work

Deformable Models. 3D bones reconstruction from X-ray images is mostly done by SSM [1,5,6,17] for bones surface modeling, or SSIM [7,8] for further modeling bones interior density. We refer the readers to [9,22] for comprehensive overviews of the existing methods. The basic principle is to rigidly align a collection of 3D models and to characterize their non-rigid mutual principal components. Then, given one or more X-ray images, a 3D reconstruction of the bones is achieved by optimizing the model parameters to maximize the similarity between its rendered versions to the input X-ray images. Recently, [15] used a deep learning approach for detecting landmarks in X-ray images and triangulating them to 3D points. However, their network does not directly outputs bone reconstructions, and the detected 3D landmarks are only used to initialize a 3D deformable model. As a result, an SSM optimization is still required and takes around 1 min.

Reconstruction from Multiple Images. Several recent methods [24,26] use deep learning approaches for reconstructing a shape from single image, for predefined objects [27]. Geometrically, by using two or more images, it is possible to reconstruct a 3D surface by triangulating corresponding points, assuming the cameras' relative positions are known [11]. The relative poses of the cameras can be computed by matching points [20] or lines [2,14,28] descriptors. Several recent papers use deep learning approaches to reconstruct shapes from two and more images. 3D-R2N2 [4] and LSM [13] use RNNs to fuse feature from multiple images for reconstructing a binary voxels mask for representing 3D models. In contrast, [29] reconstructs one volume from each image and fuses them in a context-aware layer. [25] initializes a mesh from one of the views by [24] and refines it by repeatedly applying graph convolutional layers on its 3D coordinates with learned 2D features sampled from the projections of the 3D points on the multiple images. [3] uses deep network to reconstruct 3D models from simulated bi-planar X-ray images of a single spine vertebra by applying 2D convolutional layers to encode the images into a feature vector, and then decoding it to a 3D reconstruction using 3D convolutional layers. In contrast, our method uses more effective, and geometrically consistent network architecture that uses end-to-end 3D convolutional layers with skip connections, enabling faster and more accurate reconstruction of multi-class bones as we show in Sect. 4.2.

Computed Tomography from X-ray Images. Although mathematically, generation of computed tomography from few images is an ill-posed problem, a prior knowledge on the scanned objects can approximate the free parameters. X2CTGAN [30] uses an end-to-end deep learning approach for reconstructing a CT from X-ray images. [23] trains a patient-specific deep network to extract a CT volume from single X-ray image. [12] uses a deep network to reconstruct computed tomography of different mammalian species from single X-ray images. However, these approaches only estimate CT volumes, and another challenging segmentation step is required for extracting 3D reconstructions of the anatomical objects.

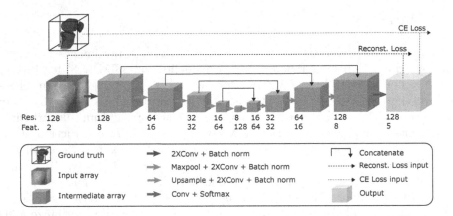

Fig. 2. Our deep network architecture and loss functions as described in Sect. 3.

3 Method

3.1 Network Architecture

Given two bi-planar X-ray images of lateral and Anterior-Posterior (AP) views, both of sizes 128×128, we first create a two channeled volume representation of size $128 \times 128 \times 128$. As illustrated in Fig.1, the volume has two channels, each contains one view (lateral or AP) replicated 128 times over one dimension (0,1 respectively). Assuming that the input images are rectified orthographic projections from orthogonal views, each axial slice in this volume contains an epipolar plane, with voxels, back-projected from pixels of two corresponding epipolar lines. Therefore, this 3D representation is geometrically consistent with the input images.

The rest of the architecture is inspired by [21] and presented in Fig. 2. We use 3D convolutions of size $3 \times 3 \times 3$, and skip connections between the encoding and the decoding layers. The last layer is a $1 \times 1 \times 1$ convolution block with 5 output channels, representing 5 output classes followed by a Softmax activation. Classes 0-4 represent an anatomical partitioning of the knee bones (see Fig. 3(e)).

3.2 Training

While CT images with ground truth 3D segmentation are available, pairs of X-ray images with associated ground truth 3D reconstructions are very rare. Moreover, geometrical alignment of each ground truth reconstruction with its X-ray images requires a 2D-3D registration process, which is itself challenging and error prone. Instead, we use annotated CT scans to create synthetic X-ray images by rendering DRRs. This way, each pair of synthetic X-ray images is associated with an aligned ground truth reconstruction.

For a supervised loss function, inspired by Fidel et al. [10] we spatially weight the cross-entropy loss to give more importance to the challenging near surface

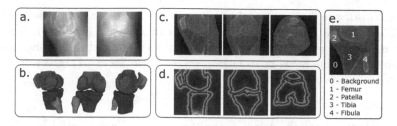

Fig. 3. Test set qualitative results for DRR input images (a-c), and training data visualization(d-e). (a) Biplanar input DRRs from the Test set. (b) Our reconstruction result from different view points. (c) Our reconstruction result displayed over a referenced CT scan, on 3 different axes. (d) Slices of Distance Weight Map (DWM) on 3 different axes. (e) Bone types and their assigned labels.

voxels. For each training sample, we define a spatial 3D Distance Weight Map (DWM) that has a size of the ground truth volume where its value on voxel i is defined by:

$$DWM(i) = 1 + \gamma \cdot exp(-d(i)/\sigma) \tag{1}$$

Where d is a distance transform that specifies for each voxel its corresponding distance from any bone surface, and γ, σ are constants which we set to $8, 10$ respectively for all the training samples. A visual example is presented in Fig. 3(d). The DWM is then applied for weighting the voxel-wised cross entropy loss as follows:

$$loss_{CE} = -\frac{1}{N} \sum_{i=1}^{N} \sum_{k=0}^{4} DWM(i) \cdot q_k(i) \cdot log(p_k(i)) \tag{2}$$

where i is the index of a voxel, N is the total number of voxels, k is the class label, $q_k(i) \in \{0, 1\}$ and $p_k(i) \in (0, 1)$ are respectively the ground truth and network prediction probabilities of voxel i being labeled k.

We further define an unsupervised reconstruction loss to align the network prediction of bones probability map with the input X-ray images. Even though the input X-ray images contain bones together with additional anatomical elements, the image gradients of the bones are quite dominant. Therefore, the input X-rays are expected to have gradients that are relatively correlated with the DRRs from the predicted bones probability map.

The reconstruction loss is defined by:

$$Loss_{reconst} = 1 - \frac{1}{2} \left(NGCC(I_{Lat}, DRR_{Lat}) + (NGCC(I_{AP}, DRR_{AP}) \right) \tag{3}$$

where NGCC is the Normalized Gradient Cross Correlation[1], I_{AP}, I_{Lat} are the input X-ray images from AP and lateral views respectively, and DRR_{AP}, DRR_{Lat} are DRRs applied on the maximum over the bones channels

---

[1] The exact definition is given in the supplementary material.

of the network prediction. This loss encourages the network to use the available information of the input images, that can actually be used in inference time, where no supervision is available. We observe that this loss improves the generalization of the network to unseen images (see Sect. 4.1). Overall our loss function is:

$$Loss = \frac{1}{2}(Loss_{reconst} + Loss_{CE}) \qquad (4)$$

For training the network, Adam optimizer was used with initial learning rate of 10^{-2}, divided by a factor of 10 every 10 epochs. We used training,validation and test sets of 188,10 and 20 scans respectively, created from knee joint CT scans with associated GT segmentations and reconstructions. Each scan was augmented by rotating it randomly with random angles of range $(-5, 5)$ and projected into 2 bi-planar DRRs which are used as synthetic input X-rays. We trained the network for 23 epochs.

Table 1. Evaluation metrics for our results given inputs of bi-planar DRRs. The results are averaged over the test set of unseen 20 scans.

	Background	Femur	Patella	Tibia	Fibula	Bones average
Chamfer (mm)	–	1.075	1.709	1.175	1.218	1.294
Dice	0.986	0.943	0.894	0.945	0.848	0.907

3.3 Domain Adaptation

X-ray images have a different appearance than DRRs. In order to apply our deep model on X-ray images, we trained a network that is based on CycleGAN [31] to transfer them to have a DRR-style appearance. During training, in each iteration the model uses two non aligned images I_{Xray} and I_{DRR} to generate two fake images: $I_{DRR \rightarrow Xray}, I_{Xray \rightarrow DRR}$. In order to generate DRR-style images which are completely aligned with the input X-ray images we use the original CycleGAN with additional content preserving loss function:

$$L_{Cont} = 1 - \frac{1}{2}\left(ZNGCC(I_{Xray \rightarrow DRR}, I_{Xray}) + (ZNGCC(I_{DRR \rightarrow Xray}, I_{DRR}))\right) \qquad (5)$$

Where ZNGCC is the Zero Normalized Gradient Cross Correlation[2]. We trained the style transfer model with training/validation sets of 370/57 pairs of bi-planar X-ray images of the knee, for 30 epochs. In the supplementary material we show visual results of the style transfer process.

---

[2] The exact definition is given in the supplementary material.

4   Experiments

4.1   DRR Inputs

We tested our method on a test set of 20 scans (see Sect. 3.2), and evaluated the results using the ground truth 3D segmentations and reconstructions. Each pair of bi-planar DRRs is used as an input to our deep network described in Sect. 3.1. For each testing sample we used the Marching Cubes algorithm[19] to extract a set of 3D bones meshes from the predicted volumetric labels. A qualitative result is presented in Fig. 3a–3c. Quantitative metrics are calculated for each bone type and presented in Table 1. Dice (higher is better) is computed over the predicted voxels maps and Chamfer (lower is better) is computed directly on the final reconstructions.

Table 2. Quantitative evaluation of real 28 test pairs of X-rays inputs, and comparisons with the femur reconstructions of [16] and [3]. The results are averaged over the test set. The patella metrics computed only on the lateral view (GT annotations for the AP view are unavailable).

	Femur SSIM [16]		[3]	Ours				
	Manual	Perturbed	Femur	Femur	Patella	Tibia	Fibula	Bones avg.
Chamfer (mm)	7.529	8.559	3.984	**1.691**	1.198	1.135	2.873	1.778
Dice	0.803	0.783	0.878	**0.948**	0.91	0.959	0.809	0.906

Table 3. Ablation study. Measuring the importance of different components of our model for real X-rays. The results metrics are averaged over the 4 reconstructed bones.

	Full	Without DWM	Lateral only	No $Loss_{reconst}$	No style transfer
Chamfer (mm)	**1.778**	1.863	2.979	1.92	6.146
Dice	**0.906**	0.901	0.844	0.892	0.742

4.2   Real X-Ray Test Cases

We evaluated 28 test cases of X-ray images. Each pair of lateral and AP X-ray images was cropped manually by an expert to contain a bi-planar pair of rectified images of the knee joint such that, when resized to 128×128 pixels, the pixel size is 1 mm. Even though the view directions of the X-ray images are not guaranteed to be exactly orthogonal, our method, trained on purely orthogonal inputs, handled such cases successfully. We applied domain adaptation procedure to transform their style as described in Sect. 3.3, and applied the network on the transformed X-ray images. In Fig. 4 we present a qualitative results of a 3D reconstruction given an input of bi-planar real X-ray images. Since 3D ground truth is not available for the X-ray images, we use 2D bi-planar ground truth multi-class masks annotated by experts for each case for evaluation: each reconstructed 3D model is projected to the 2 X-ray views and the evaluation metrics are computed relative to the GT masks.

We compare our performance with two baseline methods: the femur SSIM model[3] of [16], and the single bone reconstruction deep network of [3], trained using our training set for reconstructing the femur, and tested with the real X-ray test images (after applying our domain adaptation). Quantitative comparisons are presented in Table 2. Since [16] requires initialization for the SSIM model, we initialized it manually to the best of our ability. The optimization of [16] converged after 4.88 s, while our method, without any initialization reconstructs 4 bone types in 0.5 s, and achieves better results. Our method is more accurate and much faster than [3] which runs in 45 s for one bone reconstruction. To demonstrate the initialization sensitivity of [16], for each case of the test cases we applied a random perturbation on the manual initialization: we shifted the position parameters in a range of 20mm, multiplied the scale parameters by a factor of range [0.985, 1.015], and evaluated the average results (see Table 2, "Perturbed"). The average running time for the perturbed initialization increased from 4.88 s to 6.05 s, while 34% of the perturbed cases did not converge at all. We further show an ablation study in Table 3 of running the model without several of its components to evaluate their importance.

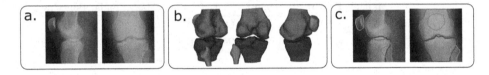

Fig. 4. Qualitative results on real X-rays. (a) Biplanar input X-rays. (b) 3D reconstruction result displayed from different view angles. (c) Boundaries of reconstruction projections displayed over the input X-rays.

Technical Details. We performed all of our experiments on a computer with MS Windows 10 64bit OS, Intel i7 7700K CPU and Nvidia GeForce GTX 1070 graphic card. The data is provided by a third party who has obtained consent for use in research.

5   Conclusion

We presented an effective end-to-end deep network for knee bones 3D reconstruction from bi-planar X-ray scans. We used a novel representation, training from synthetic data and domain adaptation to achieve an efficient, robust and accurate method. In the future we would like to extend our method to more bones reconstruction setups, and to extend the geometric 2D-3D representation of our model for additional X-ray projection models.

---

[3] Only their femur model has an available code.

Acknowledgements. The authors thank Aliza Tarakanov, Arie Rond, Noy Moskovich, Eyal Shenkman, Eitan Yeshayahu, Yara Hussein, Astar Maloul-Zamir, Liam Simani, Polina Malahov and Amit Hadari M.D, for their help in datasets creation and annotation.

References

1. Baka, N., et al.: 2D–3D shape reconstruction of the distal femur from stereo x-ray imaging using statistical shape models. Med. Image Anal. **15**(6), 840–850 (2011)
2. Ben-Artzi, G., Kasten, Y., Peleg, S., Werman, M.: Camera calibration from dynamic silhouettes using motion barcodes. In: Proceedings of the IEEE Conference on Computer Vision and Pattern Recognition, pp. 4095–4103 (2016)
3. Chen, C.-C., Fang, Y.-H.: Using bi-planar x-ray images to reconstruct the spine structure by the convolution neural network. In: Lin, K.-P., Magjarevic, R., de Carvalho, P. (eds.) ICBHI 2019. IP, vol. 74, pp. 80–85. Springer, Cham (2020). https://doi.org/10.1007/978-3-030-30636-6_11
4. Choy, C.B., Xu, D., Gwak, J.Y., Chen, K., Savarese, S.: 3D-R2N2: a unified approach for single and multi-view 3D object reconstruction. In: Leibe, B., Matas, J., Sebe, N., Welling, M. (eds.) ECCV 2016. LNCS, vol. 9912, pp. 628–644. Springer, Cham (2016). https://doi.org/10.1007/978-3-319-46484-8_38
5. Cootes, T.F., Taylor, C.J., Cooper, D.H., Graham, J.: Active shape models-their training and application. Compu. Vis. Image Underst. **61**(1), 38–59 (1995)
6. Cresson, T., Branchaud, D., Chav, R., Godbout, B., de Guise, J.A.: 3D shape reconstruction of bone from two x-ray images using 2D/3D non-rigid registration based on moving least-squares deformation. In: Medical Imaging 2010: Image Processing, vol. 7623, p. 76230F. International Society for Optics and Photonics (2010)
7. Ehlke, M., Ramm, H., Lamecker, H., Hege, H.C., Zachow, S.: Fast generation of virtual x-ray images for reconstruction of 3D anatomy. IEEE Trans. Visual. Comput. Graph. **19**(12), 2673–2682 (2013)
8. Fotsin, T.J.T., Vázquez, C., Cresson, T., De Guise, J.: Shape, pose and density statistical model for 3D reconstruction of articulated structures from x-ray images. In: 41st Annual International Conference of the IEEE Engineering in Medicine and Biology Society (EMBC), pp. 2748–2751. IEEE (2019)
9. Goswami, B., Misra, S.: 3D modeling of x-ray images: a review. Int. J. Comput. Appl **132**(7), 40–46 (2015)
10. Guerrero-Pena, F.A., Fernandez, P.D.M., Ren, T.I., Yui, M., Rothenberg, E., Cunha, A.: Multiclass weighted loss for instance segmentation of cluttered cells. In: 25th IEEE International Conference on Image Processing (ICIP), pp. 2451–2455. IEEE (2018)
11. Hartley, R., Zisserman, A.: Multiple View Geometry in Computer Vision. Cambridge University Press, Cambridge (2003)
12. Henzler, P., Rasche, V., Ropinski, T., Ritschel, T.: Single-image tomography: 3D volumes from 2D cranial x-rays. In: Computer Graphics Forum, vol. 37, pp. 377–388. Wiley Online Library (2018)
13. Kar, A., Häne, C., Malik, J.: Learning a multi-view stereo machine. In: Advances in Neural Information Processing Systems, pp. 365–376 (2017)
14. Kasten, Y., Ben-Artzi, G., Peleg, S., Werman, M.: Fundamental matrices from moving objects using line motion barcodes. In: Leibe, B., Matas, J., Sebe, N., Welling, M. (eds.) ECCV 2016. LNCS, vol. 9906, pp. 220–228. Springer, Cham (2016). https://doi.org/10.1007/978-3-319-46475-6_14

15. Kim, H., Lee, K., Lee, D., Baek, N.: 3D reconstruction of leg bones from x-ray images using CNN-based feature analysis. In: International Conference on Information and Communication Technology Convergence (ICTC), pp. 669–672. IEEE (2019)

16. Klima, O., Kleparnik, P., Spanel, M., Zemcik, P.: GP-GPU accelerated intensity-based 2D/3D registration pipeline. In: Proceedings of Shape Symposium. Swiss Institute for Computer Aided Surgery, Delemont (2015)

17. Lamecker, H., Wenckebach, T.H., Hege, H.C.: Atlas-based 3D-shape reconstruction from x-ray images. In: 18th International Conference on Pattern Recognition (ICPR 2006), vol. 1, pp. 371–374. IEEE (2006)

18. LeCun, Y., Bengio, Y., Hinton, G.: Deep learning. Nature **521**(7553), 436–444 (2015)

19. Lorensen, W.E., Cline, H.E.: Marching cubes: a high resolution 3D surface construction algorithm. ACM SIGGRAPH Comput. Graph. **21**(4), 163–169 (1987)

20. Lowe, D.G.: Distinctive image features from scale-invariant keypoints. Int. J. Comput. Vis. **60**(2), 91–110 (2004). https://doi.org/10.1023/B:VISI.0000029664.99615.94

21. Milletari, F., Navab, N., Ahmadi, S.A.: V-Net: fully convolutional neural networks for volumetric medical image segmentation. In: Fourth International Conference on 3D Vision (3DV), pp. 565–571. IEEE (2016)

22. Reyneke, C.J.F., Lüthi, M., Burdin, V., Douglas, T.S., Vetter, T., Mutsvangwa, T.E.: Review of 2-D/3-D reconstruction using statistical shape and intensity models and x-ray image synthesis: toward a unified framework. IEEE Rev. Biomed. Eng. **12**, 269–286 (2018)

23. Shen, L., Zhao, W., Xing, L.: Patient-specific reconstruction of volumetric computed tomography images from a single projection view via deep learning. Nat. Biomed. Eng. **3**(11), 880–888 (2019)

24. Wang, N., Zhang, Y., Li, Z., Fu, Y., Liu, W., Jiang, Y.G.: Pixel2Mesh: generating 3D mesh models from single RGB images. In: Proceedings of the European Conference on Computer Vision (ECCV), pp. 52–67 (2018)

25. Wen, C., Zhang, Y., Li, Z., Fu, Y.: Pixel2Mesh++: multi-view 3D mesh generation via deformation. In: Proceedings of the IEEE International Conference on Computer Vision, pp. 1042–1051 (2019)

26. Wu, J., Wang, Y., Xue, T., Sun, X., Freeman, B., Tenenbaum, J.: MarrNet: 3D shape reconstruction via 2.5 D sketches. In: Advances in Neural Information Processing Systems, pp. 540–550 (2017)

27. Wu, Z., et al.: 3D ShapeNets: a deep representation for volumetric shapes. In: Proceedings of the IEEE Conference on Computer Vision and Pattern Recognition, pp. 1912–1920 (2015)

28. Wurfl, T., Aichert, A., Maass, N., Dennerlein, F., Maier, A.: Estimating the fundamental matrix without point correspondences with application to transmission imaging. In: The IEEE International Conference on Computer Vision (ICCV), October 2019

29. Xie, H., Yao, H., Sun, X., Zhou, S., Zhang, S.: Pix2Vox: context-aware 3D reconstruction from single and multi-view images. In: Proceedings of the IEEE International Conference on Computer Vision, pp. 2690–2698 (2019)

30. Ying, X., Guo, H., Ma, K., Wu, J., Weng, Z., Zheng, Y.: X2CT-GAN: reconstructing CT from biplanar x-rays with generative adversarial networks. In: Proceedings of the IEEE Conference on Computer Vision and Pattern Recognition, pp. 10619–10628 (2019)
31. Zhu, J.Y., Park, T., Isola, P., Efros, A.A.: Unpaired image-to-image translation using cycle-consistent adversarial networks. In: Proceedings of the IEEE International Conference on Computer Vision, pp. 2223–2232 (2017)

Cellular/Vascular Reconstruction Using a Deep CNN for Semantic Image Preprocessing and Explicit Segmentation

Leila Saadatifard[(⊠)], Aryan Mobiny, Pavel Govyadinov, Hien Van Nguyen, and David Mayerich

Department of Electrical and Computer Engineering, University of Houston, Houston, USA
lsaadatifard@uh.edu

Abstract. Maps of brain microarchitecture are important for understanding neurological function, behavior, and changes due to chronic conditions, such as neurodegenerative diseases. New high-throughput microscopy techniques produce whole organ data sets imaged at subcellular resolution. The resulting volumetric data is composed of densely packed cells and interconnected microvascular networks. The data size and complexity makes manual annotation impractical and automatic segmentation challenging. In this paper, we propose a processing pipeline to segment, reconstruct, and analyze cellular and vascular microstructures in large rodent brain volumes. We first introduce a fully-convolutional, deep, and densely-connected encoder-decoder for pixel-wise semantic segmentation as a pre-processing step in our pipeline. Excessive memory complexity is mitigated by compressing the features passed through skip connections, resulting in fewer parameters and enabling a significant performance increase over prior architectures. We then quantify the pipeline's scalability, accuracy, and reliability for extracting explicit cellular and vascular images, including vessel connectivity. Finally, we demonstrate the viability of this processing pipeline on a large $(1\,\mathrm{mm}^3)$ region of the mouse somatosensory cortex as a proof of efficacy.

Keywords: Cell localization · Semantic segmentation · Vessel tracking · Whole brain structure

1 Introduction

Structural relationships between cells and microvessels are critical to brain function [4] and both are frequent targets in neurodegenerative research [12]. The regional density of cells and their distributions relative to the vascular network is therefore key to studying brain hemodynamics and neurovascular architecture in both diseased and healthy tissue [16,27]. However, disease-induced microstructural changes occur across large tissue regions, making comprehensive studies difficult due to limitations in microscope fields of view ($<1\,\mathrm{mm}$) and imaging speed.

© Springer Nature Switzerland AG 2020
F. Deeba et al. (Eds.): MLMIR 2020, LNCS 12450, pp. 134–144, 2020.
https://doi.org/10.1007/978-3-030-61598-7_13

High-throughput microscopy [19,20] aims to produce three-dimensional images of organ-scale tissue samples ($>1\,\mathrm{cm}^3$) at sub-cellular resolution ($<1\,\mu\mathrm{m}^3$). These images have sub-micrometer spatial resolution and are densely packed with a variety of microstructures. These features make manual annotation impractical and confound automated segmentation. Moreover, large-scale microscopy results in terabyte-scale volumetric data sets. Data at this scale is particularly challenging to manage and process, due to memory limitations and the computational complexity of current algorithms [2]. This indicates a compelling need for a fast and efficient frameworks that precisely map large imaces of Feulgen-stained three-dimensional samples, which would make high-throughput microscopy practical for quantitative biomedical studies.

Traditional cell segmentation methods such as Laplacian of Gaussian [17], watershed [30], and graph-cut [3] algorithms are prone to oversegmentation, particularly since cell nuclei are heterogeneous at high resolution. Recent work on microvascular modeling has focused on light-sheet and multi-photon microscopy [5,9,18]. These methods can achieve voxel sizes of $1\,\mu\mathrm{m}$–$3\,\mu\mathrm{m}$, allowing reconstruction of arterioles and venules, but are not able to precisely model the holistic 3D structure of the whole brain due to limited resolution or time constraints. In microscopy data, heterogeneous cells and low-contrast images are the challenges that state of the art fail in detecting cells with high accuracy and speed [25,26]. Prior work in vessel segmentation uses a combined template-matching and predictor/corrector approach [8]. These algorithms are less accurate with nuclear stains [29], which have lower contrast than vascular perfusion labels and exhibit additional structural complexity from chromatin features in the nucleus. Our analysis shows that introducing a deep semantic segmentation step provides a significant improvement in model accuracy over existing algorithms.

In this paper, we propose a fast and robust cell and microvascular modelling framework to quantify tissue microstructure in high-throughput three-dimensional microscopy. We establish two processing stages (Fig. 1). First, we design a memory-efficient deep neural network, named DVNet, as a preliminary semantic segmentation step to identify cellular and vascular components in thionine-stained images. DVNet takes as input raw low-contrast volumetric data and performs a pixel-wise classification. The network then outputs a mask distinguishing between cellular and vascular structures, as well as the surrounding interstitial space. A GPU-based reconstruction unit consisting of a cell localization algorithm and a vessel tracking algorithm then builds an explicit model of nuclear positions and the surrounding microvascular connectivity. The cellular and vascular structure for a $1\,\mathrm{mm}^3$ region of a thionine-stained mouse brain tissue is automatically segmented, demonstrating reliability, scalability, and accuracy exceeding prior methods.

2 Methodology

Volumetric images provide sufficient detail to quantify cellular and vascular structures and distributions. However, the low contrast and complex embedded structures makes cell localization and vascular segmentation challenging.

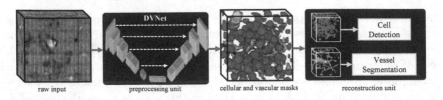

Fig. 1. Processing pipeline of the proposed framework. The input volume is a stack of low contrast images of cellular, vascular, and neuropil structure from the mouse brain. It is passed to DVNet which employs pixelwise classification, improving the contrast of the raw dataset. A GPU based postprocessing unit is applied on prediction masks to reconstruct cellular and vascular structure.

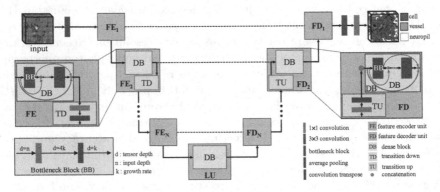

Fig. 2. Diagram of the proposed DVNet and its building blocks.

To address this issue, we designed a two-stage processing pipeline. The raw KESM input images are passed to a semantic segmentation pre-processing unit to differentiate cellular and microvascular structures. Deep neural networks (DNNs) have gained attention by outperforming state-of-the-art algorithms in many computer vision tasks including semantic segmentation, with U-Net [24], VNet [21], and DenseNet [11] frequently adopted for the medical image segmentation tasks. We develop a deep fully-convolutional network that performs semantic segmentation to classify pixels into cellular, vascular, and background components. It efficiently combines the long skip connections proposed in U-Net [24] with the short connections proposed in the DenseNet family [11] to improve the gradient back-propagation, minimize the representational information loss, and boost the overall prediction performance. Finally, the generated masks undergo reconstruction algorithms to create the cellular and vascular model. We show that the cascade of semantic segmentation step and the employed processing algorithms provides a significant improvement in the over all accuracy of the proposed framework over existing algorithms used to process the brain dataset.

2.1   Preprocessing Unit

The proposed Dense-VNet (DVNet) architecture contains densely-connected encoder and decoder paths joined by a linking unit. The encoder path decomposes the input image into a feature hierarchy, while the decoder reconstructs spatial context while performing per-pixel classification. The linking unit (LU) provides the primary information path and contains the most compact representation of the input. The cascades of feature encoder (FE) and feature decoder (FD) units run in parallel to form the encoder and decoder paths respectively (see Fig. 2).

A dense block (DB) is composed of multiple bottleneck blocks with short (also known as dense) connections from each and every layer to all subsequent layers. A bottleneck block (BB) is the main component of a dense block. Each BB contains two convolutional layers: a 1×1 convolution layer with $4k$ convolution filters followed by a 3×3 layer with k filters, where k is a hyper-parameter called growth rate [11]. Transition block is used along with a dense block to form a feature extraction unit (see Fig. 2) and aims to change the feature map sizes spatially. Transition down (TD) blocks use two convolution layers to downsample the feature maps by $2\times$ at each level of the encoder path. Transition Up (TU) blocks are used at each level of the decoder path to up-sample the feature maps by $2\times$ using two transposed-convolution layers.

Feature Compression: In DVNet, employing both the short and long skip connections together rapidly increases the number of concatenated feature maps and causes memory explosion. In Tiramisu network [14], the short skip connections of the decoder path were removed so as to decrease the number of feature maps and mitigate the explosion. This, however, leads to information loss and causes vanishing gradients which eventually degrades the convergence and prediction accuracy performance. We propose compression factors named θ_D and θ_U to control the number of feature maps in the encoder and decoder paths respectively. More specifically, if the output of a DB contains M feature maps, the following TD (TU) block will generate $\lfloor \theta_D M \rfloor$ ($\lfloor \theta_U M \rfloor$) output feature maps using a 1×1 convolution layer, where $0 < \theta_D, \theta_U \leq 1$ and $\lfloor . \rfloor$ represents the floor function.

2.2   Reconstruction Unit

The cellular and vascular masks created by the preprocessing unit are used to generate an explicit model of the brain microstructure. We adapt the state-of-the-art cell localization [26] and vascular tracing [8] algorithms to segment cell nuclei and identify vascular surfaces and connectivity (Fig. 1).

Cell Localization: The positions of different size cells in the 3D dataset are chosen using a voting algorithm (iVote3) [26]. It is an embarrassingly parallel and computationally intensive method. The implementation is accelerated using GPU shared memory and atomic operations to improve the timing performance.

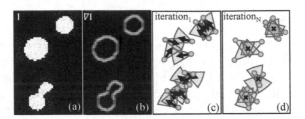

Fig. 3. iVote3 for cell detection. (a) the cell mask generated by DVNet, (b) the gradient of the input data, (c) voting regions (purple cones) for a few random pixels (green circles), with arrows showing gradient direction at each pixel, (d) voting regions get updated after each iteration; converging to the cell center in the last iteration. (Color figure online)

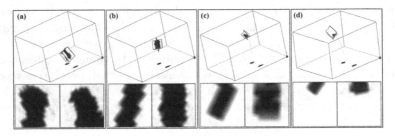

Fig. 4. Starting with some seed point the algorithm attempts to predict the direction of the fiber (a) by placing two rectangular templates perpendicular to each other (in red and blue), such that the centerline of the fiber embedded in the tissue is in the exact center (b). The algorithm then continues by advancing along the optimal direction, with an optimal size while tracking previous location along the fiber (c). The algorithm terminates when the fiber disappears (d) or when the algorithm encounters a previously segmented fiber. (Color figure online)

iVote3 computes the gradient of the input, and uses the gradient information to calculate a value for each voxel that represents the probability of being a cell center. Based on the probability mask, the gradient information is refined and a new mask is computed iteratively, until the algorithm converges and outputs a center point for each cell. Figure 3 represents an overview of iVote3. This algorithm is fast, requires minimal tuning and shows superior cell detection results on dense and low contrast datasets.

Centerline Segmentation: The segmentation is handled by a predictor-corrector algorithm that leverages texture sampling to extract centerline, radius and topology information of the microvascular network. The algorithm utilizes a set of 2 rectangular templates perpendicular to one another in order to estimate the centerline using multiple steps [8]. The process is illustrated in Fig. 4. The final output is a connected graph $G = [\mathbf{V}, \mathbf{E}]$, where v_i is the location of a branch point and e_i is a micro-vessel connecting two branch points.

3   Experiments and Results

3.1   Training Procedure

All models are trained to minimize the pixel-wise cross-entropy loss using ADAM [15] optimizer. We started the training with an initial learning rate of 0.001 and reduced it with a decay rate of 0.97 every 500 iterations following a step-wise approach. All model parameters are initialized using Xavier initialization [7]. The pixel-wise accuracy and the intersection over union (IoU) are used to evaluate the performance of the trained network. The final architectures are the result of random search over hyper-parameters such as the number of feature extraction levels, number of bottleneck blocks, and growth rate. The batch size was set to 16 and 2 for the 2D and 3D networks respectively. We evaluated the validation accuracy after every 500 iterations and saved the model with the best validation performance. DVNet is trained end-to-end using TensorFlow (version 1.13.1) [1]. We will release the source code and trained models for public evaluation upon publication.

3.2   Dataset

Data Description. We use a volumetric dataset of the thionine-stained mouse cortex imaged by KESM. The dataset was cropped into tissue sections with a 1 mm thickness and each section was imaged at 0.6 mm resolution [20]. Thionine staining is used for labeling DNA and ribosomal RNA and provides multiple structural features in a single channel. This label provides contrast for neurons, endothelial cells, and various glial cells.

Data Preparation. Three disjoint volumes of the 3D dataset are randomly selected for training and evaluation: 20 volumes of size $128 \times 128 \times 64$ voxels for training, and 2 sets of 6 volumes of size $64 \times 64 \times 32$ voxels for validation and testing, with no overlapping area between the sets. All volumes are annotated manually to segment cellular and vascular structures. Real-time data augmentation is applied during training to mitigate over-fitting and improve the generalization of the model [22]. Training images of size $64 \times 64 \times 32$ are randomly cropped from the training sets, and randomly rotated (with a probability of 0.5) along the axes.

3.3   Results

We trained 2D and 3D implementations of DVNet on the 2D and 3D slices extracted from the prepared datasets. After the random search over the hyper-parameters, we propose three DVNet configurations with different memory requirements and complexity: DVNet-v1 (with $\theta_D = 0.5$, $\theta_U = 0.3$ and $k = 8$), DVNet-v2 (with $\theta_D = 0.3$, $\theta_U = 0.3$ and $k = 16$) and DVNet-v3 (with $\theta_D = 0.5$, $\theta_U = 0.3$ and $k = 16$). All DVNet models consist of 5 feature extraction levels (FEs and FDs); each of which composed of increasing numbers of bottleneck blocks $(4, 6, 8, 10, 12)$. The linking unit has 16 convolutional layers. The segmentation performance of the proposed models is compared with the state-of-the-art architectures in Table 1. Quantitative results confirm the superior performance of the DVNet architecture in comparison to the other models, with the 3D networks significantly outperforming the 2D networks. The best performing network (DVNet-v3) improves the mean IoU by 1.4% compared to Tiramisu-103 while it requires about 58% less trainable parameters.

Table 1. Comparing the segmentation performance of the proposed and state-of-the-art models on the test data set.

	Model	# params (M)	IoU (tissue)	IoU (cell)	IoU (vessel)	Mean IoU	Pixel-wise accuracy
2D	UNet [24]	31	90.5	52.8	61.1	68.2	91.5
	Tiramisu-103 [14]	9.3	90.9	53.7	62.4	69	91.8
	DVNet-v3	5.3	**91.1**	**53.8**	**63.8**	**69.5**	**92**
3D	V-Net [21]	14	92.3	62.3	61	71.9	93.1
	Tiramisu-67 [14]	9.5	92.5	63.2	59.6	71.8	93.3
	Tiramisu-103 [14]	26	92.7	65.2	60.9	73	93.5
	DVNet-v1	2.8	**92.9**	64.4	63.8	73.7	93.6
	DVNet-v2	9.2	92.8	64.3	64.7	74	93.7
	DVNet-v3	10.8	**92.9**	**65.6**	**64.8**	**74.4**	**93.9**

The trained DVNet-v3 is utilized to segment cellular and vascular structure over a large region of the mouse brain images. A $1\,mm^3$ volume of this data set covering $1600 \times 1500 \times 1000$ voxels is broken into overlapping crops of size $128 \times 128 \times 128$ voxels. The generated masks for each crop are stitched together for the neighboring regions to model the region of interest (see Fig. 5(a)). Then iVote3 and centerline segmentation are applied on the final masks to localize the cell positions and compute the centerline and radii of the vascular network. The experimental results presented in Fig. 5(b) indicates that employing the semantic pre-processing step significantly improves the precision-recall curve at all thresholds for the cell localization, and increases the vessel-tracking F-score from 0.78 to 0.92.

Finally, the segmented cellular and vascular structures are used to compute their distribution and interaction in the mouse cortex. Regional densities of cells and vessels, and the distribution of cells relative to the microvascular network are fundamental metrics that are important to understand the brain metabolism and functions [16]. The average density for all cells in the cortex area, the total vasculature of the cortical volume and the mean distance of a cell to the closest capillary are computed. Results demonstrate 2.7% of the cortical area is filled by the blood vessels and there are about 176 thousands cells in $1\,\text{mm}^3$ of the cortex, where cells are located in $14\,\text{mm}$ apart of their closest capillary. Table 2 illustrates these results are within the range of values that previously reported in the literature. Other than these metrics, we calculated the cell sizes and mean cell distance to the closest cell to quantify the regional cell distribution in the cortex area. The results verify the variety of cell sizes in the cortex area that are measured in the range of $1.1\,\mu m$–$12\,\mu m$ for cell radii and the mean value of $6.7\,\mu m$. The mean distance of a cell to the closest cell is $15.5\,\mu m$. As it is expected, most of the cortex is filled with neuropil while the cellular/vascular structure occupy about 10% of the cortex area.

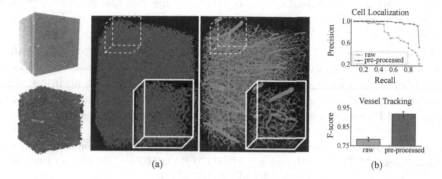

Fig. 5. (a) The pre-processed cellular (red) and vascular (green) structure for $1\,mm^3$ region of the nissl stained mouse brain by DVNet, (b) Comparing the performance of cell detection and vessel segmentation with and without semantic segmentation pre-processing. (Color figure online)

Table 2. Quantitative comparison of the measurements over the cortical regions of the mouse brain.

	Ours	Tsai et al. [27]	Wu et al. [28]	Other works
The vasculature fraction	2.7%	1%	2.6%	3.64% [10]
Cell soma distance to the clossest microvessel	$14\,\mu m$	$15\,\mu m$	$11\,\mu m$–$15\,\mu m$	
Cell density per mm^3	1.76×10^5	1.6×10^5	1.5×10^5	2.5×10^5 [6] 2.6×10^5 [13] 1.27×10^5 [23]

4   Conclusion

This paper proposes a scalable and accurate pipeline to segment, reconstruct, and analyze tissue structures composed of cellular and microvascular components. Significant gains in accuracy were obtained by leveraging a novel fully-convolutional deep neural network as a refinement step. The proposed DVNet architecture prioritizes memory efficiency by minimizing trainable parameters to learn semantic features. In particular, GPU memory limitations were significantly reduced using the proposed feature compression method, which increases performance by maximizing the information flow through the skip connections while minimizing the number of trainable parameters. The final nuclear and vascular model is explicitly reconstructed using current state-of-the-art algorithms in cell localization and vessel tracking. Explicit segmentation is significantly improved by DVNet semantic pre-processing. This pipeline enables reporting the distribution of cells and vessels in a big region of the mouse somatosensory cortex.

Future modifications to our pipeline could include cell and vessel classification. A DVNet-like architecture could potentially be trained to classify cell types with different morphologies, such as neurons and glial cells. In addition, other tissue features such as neural connectivity, fatty deposits, and collagen or muscle fibers could be semantically identified prior to explicit modeling.

Acknowledgment. This work was funded in part by the National Institutes of Health/National Library of Medicine #4 R00 LM011390-02 and the Cancer Prevention and Research Institute of Texas (CPRIT) #RR140013.

References

1. Abadi, M., et al.: Tensorflow: a system for large-scale machine learning. In: 12th {$USENIX$} Symposium on Operating Systems Design and Implementation ({$OSDI$} 2016), pp. 265–283 (2016)
2. Akil, H., Martone, M.E., Van Essen, D.C.: Challenges and opportunities in mining neuroscience data. Science **331**(6018), 708–712 (2011)
3. Al-Kofahi, Y., Lassoued, W., Lee, W., Roysam, B.: Improved automatic detection and segmentation of cell nuclei in histopathology images. IEEE Trans. Biomed. Eng. **57**(4), 841–852 (2010)
4. Andreone, B.J., Lacoste, B., Gu, C.: Neuronal and vascular interactions. Ann. Rev. Neurosci. **38**, 25–46 (2015)
5. Blinder, P., Tsai, P.S., Kaufhold, J.P., Knutsen, P.M., Suhl, H., Kleinfeld, D.: The cortical angiome: an interconnected vascular network with noncolumnar patterns of blood flow. Nat. Neurosci. **16**(7), 889 (2013)
6. Erö, C., Gewaltig, M.O., Keller, D., Markram, H.: A cell atlas for the mouse brain. Front. Neuroinform. **12**, 84 (2018). https://doi.org/10.3389/fninf.2018.00084
7. Glorot, X., Bengio, Y.: Understanding the difficulty of training deep feedforward neural networks. In: Proceedings of the Thirteenth International Conference on Artificial Intelligence and Statistics, pp. 249–256 (2010)
8. Govyadinov, P.A., Womack, T., Chen, G., Mayerich, D., Eriksen, J.: Robust tracing and visualization of heterogeneous microvascular networks. IEEE Trans. Vis. Comput. Graph. **1**, 1–1 (2018)
9. Haft-Javaherian, M., Fang, L., Muse, V., Schaffer, C.B., Nishimura, N., Sabuncu, M.R.: Deep convolutional neural networks for segmenting 3D in vivo multiphoton images of vasculature in Alzheimer disease mouse models. PloS One **14**(3), e0213539 (2019)

10. Heinzer, S., et al.: Hierarchical microimaging for multiscale analysis of large vascular networks. Neuroimage **32**(2), 626–636 (2006)

11. Huang, G., Liu, Z., Van Der Maaten, L., Weinberger, K.Q.: Densely connected convolutional networks. In: Proceedings of the IEEE Conference on Computer Vision and Pattern Recognition, pp. 4700–4708 (2017)

12. Iadecola, C.: Neurovascular regulation in the normal brain and in Alzheimer's disease. Nat. Rev. Neurosci. **5**(5), 347 (2004)

13. Irintchev, A., Rollenhagen, A., Troncoso, E., Kiss, J.Z., Schachner, M.: Structural and functional aberrations in the cerebral cortex of tenascin-C deficient mice. Cereb. Cortex **15**(7), 950–962 (2004). https://doi.org/10.1093/cercor/bhh195

14. Jégou, S., Drozdzal, M., Vazquez, D., Romero, A., Bengio, Y.: The one hundred layers tiramisu: fully convolutional densenets for semantic segmentation. In: 2017 IEEE Conference on Computer Vision and Pattern Recognition Workshops (CVPRW), pp. 1175–1183. IEEE (2017)

15. Kingma, D., Ba, J.: Adam: a method for stochastic optimization. In: International Conference on Learning Representations (2014)

16. Kleinfeld, D., et al.: A guide to delineate the logic of neurovascular signaling in the brain. Front. Neuroenergetics **3**, 1 (2011)

17. Kong, H., Akakin, H.C., Sarma, S.E.: A generalized Laplacian of Gaussian filter for blob detection and its applications. IEEE Trans. Cybern. **43**(6), 1719–1733 (2013)

18. Lauwers, F., Cassot, F., Lauwers-Cances, V., Puwanarajah, P., Duvernoy, H.: Morphometry of the human cerebral cortex microcirculation: general characteristics and space-related profiles. Neuroimage **39**(3), 936–948 (2008)

19. Li, A., et al.: Micro-optical sectioning tomography to obtain a high-resolution atlas of the mouse brain. Science **330**(6009), 1404–1408 (2010)

20. Mayerich, D., Abbott, L., McCormick, B.: Knife-edge scanning microscopy for imaging and reconstruction of three-dimensional anatomical structures of the mouse brain. J. Microsc. **231**(1), 134–143 (2008)

21. Milletari, F., Navab, N., Ahmadi, S.A.: V-Net: fully convolutional neural networks for volumetric medical image segmentation. In: 2016 Fourth International Conference on 3D Vision (3DV), pp. 565–571. IEEE (2016)

22. Mobiny, A., Singh, A., Van Nguyen, H.: Risk-aware machine learning classifier for skin lesion diagnosis. J. Clin. Med. **8**(8), 1241 (2019)

23. Murakami, T.C., et al.: A three-dimensional single-cell-resolution whole-brain atlas using CUBIC-X expansion microscopy and tissue clearing. Nat. Neurosci. **21**(4), 625 (2018)

24. Ronneberger, O., Fischer, P., Brox, T.: U-Net: convolutional networks for biomedical image segmentation. In: Navab, N., Hornegger, J., Wells, W.M., Frangi, A.F. (eds.) MICCAI 2015. LNCS, vol. 9351, pp. 234–241. Springer, Cham (2015). https://doi.org/10.1007/978-3-319-24574-4_28

25. Saadatifard, L., Mayerich, D.: Three dimensional parallel automated segmentation of neural soma in large KESM images of brain tissue. Microsc. Microanal. **22**(S3), 788–789 (2016)

26. Saadatifard, L., Abbott, L.C., Montier, L., Ziburkus, J., Mayerich, D.: Robustcell detection for large-scale 3D microscopy using GPU-accelerated iterative voting. Front. Neuroanat. **12**, 28 (2018)

27. Tsai, P.S., et al.: Correlations of neuronal and microvascular densities in murine cortex revealed by direct counting and colocalization of nuclei and vessels. J. Neurosci. **29**(46), 14553–14570 (2009)

28. Wu, J., et al.: 3D BrainCV: simultaneous visualization and analysis of cells and capillaries in a whole mouse brain with one-micron voxel resolution. Neuroimage **87**, 199–208 (2014)
29. Xiong, B., et al.: Precise cerebral vascular atlas in stereotaxic coordinates of whole mouse brain. Front. Neuroanat. **11**, 128 (2017)
30. Zhang, M., Zhang, L., Cheng, H.D.: A neutrosophic approach to image segmentation based on watershed method. Sign. Process. **90**(5), 1510–1517 (2010)

Improving PET-CT Image Segmentation via Deep Multi-modality Data Augmentation

Kaiyi Cao[1], Lei Bi[1(✉)], Dagan Feng[1,2], and Jinman Kim[1]

[1] School of Computer Science, University of Sydney, Sydney, NSW, Australia
lei.bi@sydney.edu.au

[2] Med-X Research Institute, Shanghai Jiao Tong University, Shanghai, China

Abstract. Positron emission tomography (PET) - computed tomography (CT) is a widely-accepted imaging modality for staging, diagnosis and treatment response monitoring of cancers. Deep learning based computer aided diagnosis systems have achieved high accuracy on tumor segmentation on PET-CT images in recent years. PET images can be used to detect functional structures such as tumors, whilst CT images provide complementary anatomical information. As for tumor detection using deep learning methods, multi-modality segmentation was verified to be effective. In this work, we propose a generative adversarial network (GAN) based augmentation method to synthesized multi-modality data pairs on PET and CT to improve the training of multi-modality segmentation method. Our novelty lies in creating a semantic label augmentation method to provide latent information that is suitable for the multi-modality synthesis. In addition, we set out a 'Split U' structure which can generate both PET-CT modalities from a latent input. Our experimental results demonstrated that the synthesized images generated by our method can be used to augment the training data for PET-CT segmentation.

Keywords: Data augmentation · Multi-modality · Segmentation

1 Introduction

Positron emission tomography – computed tomography (PET-CT) is regarded as the imaging modality of choice for staging, diagnosis and monitoring treatment responses of many cancer diseases [1]. For PET-CT images, there has been great research interest in the developing of automated computer aided diagnosis (CAD) system that can assist physicians in image interpretation [2]. Automatic tumor segmentation is a fundamental requirement for automated CAD system.

Compared with traditional segmentation methods such as thresholding [3], region growing [4], and region splitting/merging [5], deep learning based segmentation methods for its capability to leverage large datasets to derive high-level semantic features are regarded as the state-of-the-art. However, there is a scarcity of annotated medical image training data due to expensive data acquisition procedure and complex data annotation process [6]. Consequently, without sufficient annotated image training data to cover all the feature variations e.g., texture, shape and size, deep learning based segmentation methods have difficulties in the segmentation of e.g., tumors.

© Springer Nature Switzerland AG 2020
F. Deeba et al. (Eds.): MLMIR 2020, LNCS 12450, pp. 145–152, 2020.
https://doi.org/10.1007/978-3-030-61598-7_14

Data augmentation is an effective approach to expand data volume. Conventional data augmentation methods include random flipping, cropping, rotating shifting, scaling, and adding noise [7]. However, these methods only increase the data volume without adding additional feature variations, such as semantic variation. Recently, data augmentation based on image synthesis was introduced to overcome the limitations of insufficient feature variations [8]. The underlying concept is to derive synthetic images from limited annotated image training data. The derived synthetic images will be augmented to the existing training data to add new feature variations.

Unfortunately, existing image synthesis based data augmentation methods, only focused on a single modality transfer e.g., CT to PET [9], and did. Consequently, these existing single-modality image synthesis methods cannot be directly applied to multi-modality PET-CT images where there is the need to use information from both the modalities.

In this work, we propose a multi-modality image synthesis method for data augmentation of PET-CT images. We leveraged the state-of-the-art feature learning method of generative adversarial network (GAN) to jointly learn complementary PET-CT image features. The jointly learned PET-CT image features were then used to derive additional synthetic feature variations that can be used to augment the existing training data. When compared to existing augmentation methods, we introduce the following contributions:

We proposed a multi-modality image synthesis method. Compared to existing image synthesis method focused on single-modality, our method leverages image features form complementary imaging modality. The ability to leverage the complementary imaging modality allows to derive more realistic multi-modality synthetic images.

We demonstrated that our multi-modality synthetic images argument the training dataset, which thereby improving the overall PET-CT tumor segmentation performance without using additional annotated medical images.

2 Methods

2.1 Pre-processing

PET images were initially normalized into PET/CT standard uptake value (SUV) [10], which is defined as:

$$SUV = \frac{C_{pet(t)} body_{mass}}{injected_{dose}} \quad (1)$$

where $C_{pet(t)}$ is the pixel radioactivity concentration corrected at the injection time, and $injected_{dose}$ is the measured dose of FDG. Next, PET and CT slices which contain tumor regions are extracted and saved in pairs with a threshold cutting off top 5% largest values for PET images and 10% largest values, 20% smallest values for CT images to enhance the interested pixel region of patient body parts. After that, the pixel values are normalized to (0,255) and contrast limited adaptive histogram equalization (CLAHE) [11–13] is applied to enhance the contrast.

2.2 Semantic Label Extraction and Integration

Two kinds of semantic labels were used in this work, the ground truth labels and the augmented label. The first one is fed to the model while training as the input and real PET/CT pairs are used to constraint the output synthesized pairs. The second one, the augmented labels, are used as the input to generate the synthesized PET/CT pairs and theses paired data are fed for segmentation. The ground truth label is the combination of the regions of skin, bones and tumor labels. The regions of skin and bones can be easily obtained through thresholding the CT images. The augmented label is the same as the ground truth label except that it replaces real tumor labels with proceeded synthetic tumors. Through using a random combination of traditional methods including rotating, resizing, and random distortion [8, 14, 15], synthetic tumor regions are created. To avoid the situation where the created tumor region overlaps with bone areas, the resizing ratio is set to (0.85,1) so that the region is always smaller than the real case. Besides, the rotation angle is set within 3 degree for the same reason. After such label is created, it is combined with the skin and bone regions and used to generate paired PET/CT synthesized images.

2.3 Multi-modality Co-learning from Semantic Labels

Generative adversarial nets (GAN) [16] consists of two parts: the generator (G) and the discriminator (D). G learns the data distribution and D calibrates G through returning a cross entropy loss [17] judging whether the output from G obeys a real distribution. Two components are cultivated together to reach a Nash equilibrium that the images generated by G is so real so that D can hardly distinguish it from the real one.

Therefore, the objective of training is to minimize the loss function in Eq. (2):

$$\min_{G} \max_{D} V(D, G) = E_{x \sim p_{data}(x)}\left[\log D(x)\right] + E_{z \sim p_z}\left[\log(1 - D(G(z)))\right] \quad (2)$$

where $D(x)$ is the possibility of the real data is true judged by the discriminator and $G(z)$ is the synthesized image from a noise signal z. Through adding constraints and feeding them into the discriminator together with synthesized image, as shown in Eq. (3):

$$\min_{G} \max_{D} V(D, G) = E_{x \sim p_{data}(x)}\left[\log D(x|y)\right] + E_{z \sim p_z}\left[\log(1 - D(G(z|y)))\right] \quad (3)$$

The latent/semantic vector can also be used as a condition y to guide the synthesis result.

Our method consists of a two-branch generator, shown in Fig. 1(b) and two isolated discriminators, shown in Fig. 1(c). Using on 80% real data, we trained the model and created data of the same volume with the augmented semantic label in Sect. 2.2. We choose the rotation angle, random deformation level and size variation of the tumor label so that it can fit well into the normal body region. During the training process, $L1$ loss is used in both generator branches and cross-entropy loss is used in the discriminator. The training process is to minimize the objective:

$$\min_{G} \max_{D} V(D, G) = E_{x \sim p_{data_{PET}}(x)}\left[\log D_1(x|y)\right] + E_{z \sim p_{z_{PET}}}\left[\log(1 - D_1(G_1(z|y)))\right]$$
$$+ E_{x \sim p_{data_{CT}}(x)}\left[\log D_2((x|y))\right] + E_{z \sim p_{z_{CT}}}\left[\log(1 - D_2(G_2(z|y)))\right]$$

$$(4)$$

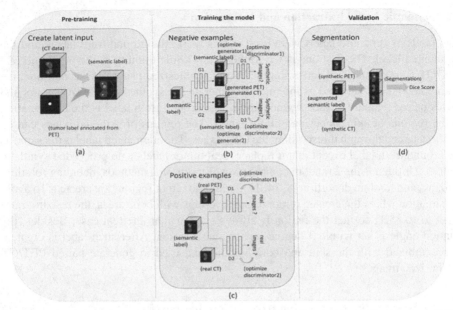

Fig. 1. Our proposed experiment procedure: (a) Create latent input; (b) (c) Training our GAN model; (d) Segmentation

The gradients from generator ($G1$ and $G2$) branches are back propagated alternately so that the model is cultivated on features of both modalities.

2.4 Experimental Setup

We used the PET-CT soft-tissue sarcoma (STS) dataset [18, 19]. It consists of 51 PET-CT scans derived from 51 patients. with the diagnosis was confirmed through histology. Only the axial slices with the presence of tumors were used in the experiment. The expert annotated tumor regions were used as the ground truth. The STS dataset was split into 5 folds, each fold contained 10 to 11 patients. All training processes used 80/20 training-testing ratio and the final results were averaged across the 5-fold cross-validation. All experiments were carried out on a NVIDIA 1080 TI GPU. The experiments consist of two stages: (1) multi-modality image synthesis; and (2) segmentation using the synthetic images. The multi-modality image synthesis model was trained for 100 epochs using an Adam optimizer with learning rate of 2×10^{-4}.

For the first stage of the experiments, we used Pix2Pix [20] as the baseline. we created three types of data pairs, which are: (i) PET-Semantic Label (PET-SL); (ii) CT-Semantic Label (CT-SL); and (iii) PET-CT-Semantic Label (PET-CT-SL). As for the baseline, we trained two separate Pix2Pix models separately on the PET-SL data pairs and the CT-SL data pairs. For our method, we trained the model on PET-CT-SL data pairs. The commonly used image synthesis matrix were used for comparison, which consists of: peak signal-to-noise ratio (PSNR), structural similarity (SSIM) and mean-square error (MSE) [21].

For the second stage experiment, a co-learning PET-CT segmentation method was used as the baseline method [22]. Co-learning method was designed for PET-CT tumor segmentations by iteratively fusing the PET and CT image features. 4 different scenarios were chosen for evaluating the effectiveness of the derived multi-modality synthetic images: (i) 100% of synthetic images; (ii) 50% of real images; (iii) 50% of synthetic images + 50% of real images; and (iv) 100% of real images. We expect a growing trend in dice scores, if the multi-modality synthetic images are useful for data augmentation.

3 Results and Discussion

3.1 Image Synthesis Analysis

Table 1 shows that our method improved PSNR by 4.6, SSIM by 7% and MAE by 61% on synthetic CT images. For PET, the PSNR increased by 3.9, SSIM increased by 4.2% and MAE is lowered by 59.9%. Figure 2 shows three example multi-modality image synthesis results and the last column (Augmented Ground Truth) is the corresponding augmented tumor labels deformed from real ones. The results in Table 1 and Fig. 2 show that the synthetic images from our method are closer to the real images, where the muscular textures in CT and tumor textures in PET are clearer.

Table 1. Comparison on CT/PET image qualities using Pix2Pix and our method in terms of PSNR, SSIM, and MSE

Modality	Method	PSNR	SSIM	MSE
CT	Pix2pix	29.03	0.869	93.930
	Our method	32.923	0.911	37.735
PET	Pix2pix	24.706	0.828	276.299
	Our method	29.366	0.898	107.742

In Fig. 3, the example of synthetic image results explains how the synthetic data leads to more meaningful latent information for multi-modality learning. Compared with the traditional method, the regions of interest (areas highlighted red) show clearer tumor contour and shape similarity compared with the ground truth label. We suggest that the two-channel learning process helps to create better understanding on semantic knowledge and leads to the coherence of synthetic images across modalities.

3.2 Segmentation

Table 2 shows that, compared with the synthesized data by a single modality synthesis method (Pix2Pix), our method improved Dice by 1.23% with 100% synthetic data; and 1.25% improvement with 50% of real data plus 50% of synthetic data. The result shows,

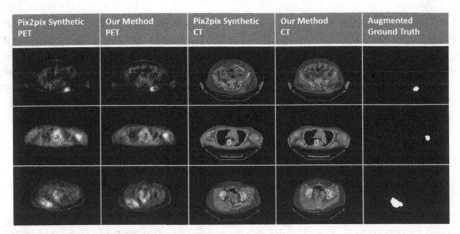

Fig. 2. Comparison on synthetic images with pix2pix and our method

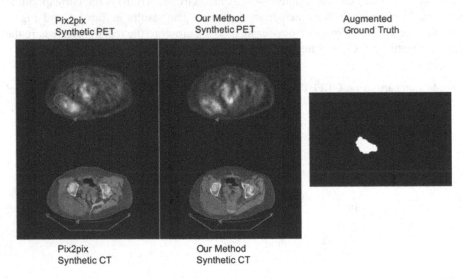

Fig. 3. Case to show the detailed synthetic difference between pix2pix and our method (Color figure online)

when the PET-CT data pairs are bound together to the semantic labels, the coherence between the modalities can also lead to better semantic understanding from the model, and further result in higher segmentation accuracies compared with single-modality methods. Figure 4 shows example segmentation results. We found that synthetic images derived from our method helped in reducing false positives (targeted in red circles).

Table 2. Dice score evaluation of co-learning segmentation

Modality	100% Synthetic	50% Real	50% Real + 50% Synthetic	100% Real
Pix2Pix	57.54%	NA	60.38%	NA
Our method	58.77%	NA	61.63%	NA
Real data	NA	60.39%	NA	62.475%

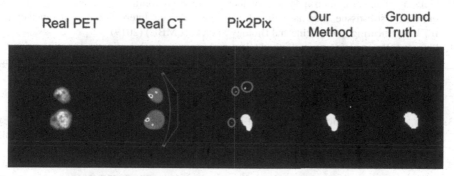

Fig. 4. Example segmentation results. Augmented synthetic data was derived from by Pix2Pix and our method. (Color figure online)

4 Conclusion

In this work, we proposed a GAN based multi-modality data augmentation method for improving PET-CY segmentation. Our results with clinical PET-CT sarcoma studies demonstrated that our method can produce useful PET-CT synthetic images and can be used to improve the existing segmentation methods.

References

1. Kratochwil, C., Haberkorn, U., Giesel, F.L.: PET/CT for diagnostics and therapy stratification of lung cancer. Der Radiologe **50**(8), 684–691 (2010)
2. Verma, B., Zakos, J.: A computer-aided diagnosis system for digital mammograms based on fuzzy-neural and feature extraction techniques. IEEE Trans. Inf Technol. Biomed. **5**(1), 46–54 (2001)
3. Fan, J.-L., Zhao, F.: Two-dimensional Otsu's curve thresholding segmentation method for gray-level images. Acta Electronica Sinica **35**(4), 751 (2007)
4. Tang, J.: A color image segmentation algorithm based on region growing. In: 2010 2nd International Conference on Computer Engineering and Technology. IEEE (2010)
5. Hu, G.: Survey of recent volumetric medical image segmentation techniques. In: Biomedical Engineering. IntechOpen (2009)
6. Ker, J., et al.: Deep learning applications in medical image analysis. IEEE Access **6**, 9375–9389 (2018)

7. Mikołajczyk, A., Grochowski, M.: Data augmentation for improving deep learning in image classification problem. In: 2018 International Interdisciplinary PhD workshop (IIPhDW). IEEE (2018)

8. Wang, J., Perez, L.: The effectiveness of data augmentation in image classification using deep learning. Convolutional Neural Netw. Vis. Recogn. **11**, 1–8 (2017)

9. Bi, L., Kim, J., Kumar, A., Feng, D., Fulham, M.: Synthesis of positron emission tomography (PET) Images via multi-channel generative adversarial networks (GANs). In: Cardoso, M.J., et al. (eds.) CMMI/SWITCH/RAMBO -2017. LNCS, vol. 10555, pp. 43–51. Springer, Cham (2017). https://doi.org/10.1007/978-3-319-67564-0_5

10. Peng, Y., et al. Deep multi-modality collaborative learning for distant metastases predication in PET-CT soft-tissue sarcoma studies. In: 2019 41st Annual International Conference of the IEEE Engineering in Medicine and Biology Society (EMBC) (2019)

11. Pisano, E.D., et al.: Contrast limited adaptive histogram equalization image processing to improve the detection of simulated spiculations in dense mammograms. J. Digit. Imaging **11**(4), 193 (1998). https://doi.org/10.1007/BF03178082

12. Pizer, S.M., et al.: Contrast-limited adaptive histogram equalization: speed and effectiveness. In: Proceedings of the First Conference on Visualization in Biomedical Computing (1990)

13. Reza, A.M.: Realization of the contrast limited adaptive histogram equalization (CLAHE) for real-time image enhancement. J. VLSI Sign. Process. Syst. Sign. Image Video Technol. **38**(1), 35–44 (2004). https://doi.org/10.1023/B:VLSI.0000028532.53893.82

14. Um, T.T., et al.: Data augmentation of wearable sensor data for Parkinson's disease monitoring using convolutional neural networks. arXiv preprint arXiv:1706.00527 (2017)

15. Taylor, L., Nitschke, G.: Improving deep learning using generic data augmentation. arXiv preprint arXiv:1708.06020 (2017)

16. Goodfellow, I., et al.: Generative adversarial nets. In: Advances in Neural Information Processing Systems (2014)

17. Panchapagesan, S., et al.: Multi-task learning and weighted cross-entropy for DNN-based keyword spotting. In: INTERSPEECH (2016)

18. Vallières, M., et al.: A radiomics model from joint FDG-PET and MRI texture features for the prediction of lung metastases in soft-tissue sarcomas of the extremities. Phys. Med. Biol. **60**(14), 5471 (2015)

19. Clark, K., et al.: The cancer imaging archive (TCIA): maintaining and operating a public information repository. J. Digit. Imaging **26**(6), 1045–1057 (2013). https://doi.org/10.1007/s10278-013-9622-7

20. Isola, P., et al.: Image-to-image translation with conditional adversarial networks. In: Proceedings of the IEEE Conference on Computer Vision and Pattern Recognition (2017)

21. Hore, A., Ziou, D.: Image quality metrics: PSNR vs. SSIM. In: 2010 20th International Conference on Pattern Recognition. IEEE (2010)

22. Kumar, A., et al.: Co-learning feature fusion maps from PET-CT images of lung cancer. IEEE Trans. Med. Imaging **39**(1), 204–217 (2019)

Stain Style Transfer of Histopathology Images via Structure-Preserved Generative Learning

Hanwen Liang[1], Konstantinos N. Plataniotis[1], and Xingyu Li[2(✉)]

[1] The Edward S. Rogers Department of Electrical and Computer Engineering,
University of Toronto, Toronto, Canada
[2] Electrical and Computer Engineering, University of Alberta, Edmonton, Canada
xingyu@ualberta.ca

Abstract. Computational histopathology image diagnosis becomes increasingly popular and important. While pathologists do not struggle with color variations in slides, computational solutions usually suffer from this critical issue. In this regard, this study proposes two stain style transfer models, SSIM-GAN and DSCSI-GAN, based on the generative adversarial networks. By cooperating structural preservation metrics and feedback of an auxiliary diagnosis net in learning, medical-relevant information presented by image texture, structure, and chroma-contrast features is preserved in color-normalized images. Particularly, the smart treat of chromatic image content in our DSCSI-GAN model helps to achieve noticeable normalization improvement in image regions where stains mix due to histological substances co-localization. Extensive experimentation on public histopathology image sets indicates that our methods outperform prior arts in terms of generating more stain-consistent images, better preserving histological information in images, and obtaining significantly higher learning efficiency. Our python implementation is published on https://github.com/hanwen0529/DSCSI-GAN.

Keywords: Stain style transfer · Generative model · Color normalization · Structural similarity · Computational histopathology

1 Introduction

Computational histopathology is a promising field where image processing and machine learning techniques are applied to histopathology images for disease diagnosis. One critical issue with it is color variation among histopathology images. Due to the variation in chemical stains and staining procedures, tissue slides can differ greatly from each other in visual appearance. Other factors may also introduce variation to visual appearance, including the storage conditions

Electronic supplementary material The online version of this chapter (https://doi.org/10.1007/978-3-030-61598-7_15) contains supplementary material, which is available to authorized users.

of stain prior to use and the handling of slides. Since color information is recognized as key factor in an automatic histopathology analysis system, the tissue slices' varying appearance directly increases diagnosis complexity and impacts the quality and accuracy of a computational solution [1]. To address the color variation issue in computational histopathology, one potential, practical solution is to eliminate color variation in the pre-learning stage and many color normalization solutions are proposed. Briefly, prior color normalization approaches can be categorized into three groups: (i) histogram matching methods [2–4]; (ii) spectral matching based on stain decomposition [4–7]; (iii) style transfer based on generative learning [8–12]. Among the three categories, histogram matching methods treat color distribution independent of image content, thus may introduce image distortions after normalization. The spectral matching methods are based on Beer-Lambert law for stain decomposition [13] and achieve good performance for light-absorbing stains. However, there are many scattering stains in histopathology images that do not follow Beer-Lambert Law. To overcome these limitations, deep learning methods, especially generative adversarial networks (GAN) [14], are exploited, hoping for a generalizable color normalization solution for computational histopathology.

In this study, we develop a novel stain-style transfer framework combining a GAN network and a classification network for color normalization on histopathology images. The proposed framework learns the histopathological staining protocol from training set and achieves stain style transformation with high efficiency. To preserve histopathological information delivered by texture, structural, and color content in images, we respectively innovate the use of structural similarity index matrix (SSIM) [15] and directional statistics based color similarity index (DSCSI) [16] as the image reconstruction loss function in training process. A feature preservation loss function is exploited in the proposed framework to minimize the loss of discriminative representations in images. We perform extensive experimentation to evaluate the proposed SSIM-GAN and DSCSI-GAN models against prior arts. The results suggest that the proposed models succeed in transferring stain style, generating stain-consistent images, and bringing significantly higher learning efficiency than prior arts.

In summary, our contributions are in two-folds. First, we propose two stain style transfer models that learn histopathological staining protocols and realize color normalization. Second, we introduce the use of SSIM and DSCSI metrics in GAN's learning, preserving structural information in images when transferring color patterns. DSCSI-GAN is the first to utilize image chromatic spatial organization to regularize GAN in stain style transfer and achieves noticeable normalization improvement in image regions where chemical stains mix due to histological substances overlap.

2　Method

To clarify the histopathological stain style transfer problem, we define the dataset of histopathology images from pathology lab A as $X_A = \{x_A^1, \dots x_A^n\}$ and its corresponding labels $Y_A = \{y_A^1, \dots y_A^n\}$, where $x_A^i \in R^3$ is a color image in RGB format and $y_A^i \in (0, 1)$ indicates whether it is normal or tumor image. We also define the test dataset from pathology lab B as X_B which represents the same type of tissue but has different staining appearance. The proposed stain normalization model aims at generating images $\dot{X}_B$ from source images X_B, where $\dot{X}_B$ should preserve the histological information in X_B and have the same color style as training set X_A.

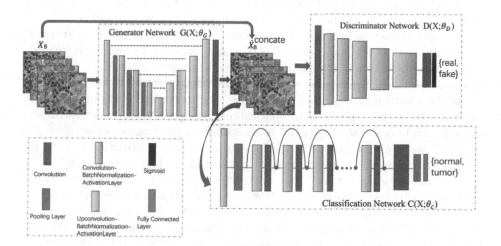

Fig. 1. Overview of the stain-style transfer network and classification network.

Network Architectures: The block diagram of the proposed framework is shown in Fig. 1. The GAN model comprises of a generator $G(X; \theta_G)$ and a discriminator $D(X; \theta_D)$, where this works adapts U-Net [17] as $G(X; \theta_G)$ to communicate image histological content between layers and to embrace precise localization property for color-normalized image generation. For discriminator $D(X; \theta_D)$, we adopt the discriminator part from vanilla GAN [14], which takes real images or fake images generated by $G(X; \theta_G)$ for classification. Our framework also deploys ResNet [18] as the auxiliary model $C(X; \theta_C)$ to classify histopathology images as tumor or normal images. After the network $C(X; \theta_C)$ is trained on X_A, we use its feature map before the last fully-connected layer as feedback for GAN's update.

The framework learns color characteristics within an entire dataset X_A in training and finally is capable of processing query images X_B so that the normalized images $\dot{X}_B$ appear the same stain style of X_A. We cooperate image structural similarity loss functions in GAN's learning for histological information preservation.

Objective Functions: Given the proposed GAN based stain transfer framework with parameter set $\theta = \{\theta_G, \theta_D, \theta_C\}$, where θ_G and θ_D are the trainable parameters of generator and discriminator of GAN and θ_C is the trainable parameters of the diagnosis net, the objective function $L(\theta)$ of the proposed model is composed of three loss functions:

$$L(\theta) = \alpha L_{GAN}(G, D) + \beta L_{reco}(G) + \gamma L_{fp}(G, C)), \tag{1}$$

where α, β, γ are the weights to balance GAN loss $L_{GAN}(G, D)$, image reconstruction loss $L_{reco}(G)$, and feature preserving loss $L_{fp}(G, C)$. The GAN loss $L_{GAN}(G, D)$ [14] drives the system to perform an adversarial game. Since the generic GAN loss $L_{GAN}(G, D)$ may induce the model to generate an image losing histological patterns, we introduce two more loss functions in training.

The reconstruction loss $L_{reco}(G)$ measures the difference between generated images $\dot{X}_A$ and its original counterpart X_A, and enforces the generator learning image color distribution and maintaining the structural information in images at the same time. Specifically, structural information refers to the knowledge about the structure of objects, e.g. spatially proximate, in the visual scene [19]. Particularly in the context of computational histopathology, structural information mainly refers to the spatial organization of histological substances, i.e. multicellular structures, in histopathology images. Such information is a key for downstream computational histopathology and thus should be maintained in color normalization. In prior arts, generative networks usually adopt MSE as image reconstruction loss function. However, MSE-driven models are prone to generating a smooth/blur reconstruction where some structural information in the original signal is missing [20]. To address this problem, we introduce two loss functions based on image structural similarity (i.e. SSIM and DSCSI) to measure quality of generated images. The motivations behind is that structural similarity correlates well with human's perception of image quality [20] and facilitates the networks to maintain the texture and structural patterns in images.

The SSIM based reconstruction loss function when training $G(X; \theta_G)$ and $D(X; \theta_D)$ can be formulated by

$$L_{reco}(G) = E_{x \sim X_A}[1 - SSIM(x, G(x; \theta_G))], \tag{2}$$

where $L_{reco}(G) \in [0, 1]$ and SSIM [15] is the structural similarity index matrix between original image x_A^i and generated image $\dot{x}_A^i$ by $G(X; \theta_G)$. As SSIM is proposed for gray-scale images, in practice, we first map RGB images to gray-scale images. A sliding window is applied to obtained gray-scale images and image differences within the sliding windows, characterized by luminance, structure, and contrast, are evaluated and averaged for a single SSIM value. We name the stain style transfer model with SSIM as SSIM-GAN in this study.

Note SSIM is proposed for grayscale image quality measurement and may fail for color images that exhibits chromatic deviations [16]. Aware that histopathology images are color in nature and color normalization in these images focuses on chromatic style transfer, we also develop a DSCSI based reconstruction loss and replace the SSIM metric in GAN learning. The corresponding model is called DSCSI-GAN in this paper. The DSCSI loss combines chromatic and achromatic similarity and is formulated as:

$$L_{reco}(G) = E_{x \sim X_A}[1 - DSCSI(x, G(x; \theta_G))]. \tag{3}$$

In training, the original image and generated image are first transformed to the S-CIELAB space [21] and six similarity measures in the hue/chroma/lightness channels are developed to compute the similarity score in SDCSI loss.

The last term in Eq. (1) is used to preserve discriminative histological features in images. In training, we feed original image x_A and generated image $\dot{x}_A$ to the pre-trained auxiliary diagnosis net $C(X; \theta_C)$ and extract feature representations after the final convolution layers. Following Cho's work [9], we obtain feature-preserving loss by means of Kullback-Leibler divergence.

Training Procedure: Our proposed methods are composed of two learning stages. First, a diagnosis net used as the auxiliary classifier $C(X; \theta_C)$ is trained on the training set X_A and image labels. Second, the adversary generating model is trained following Algorithm 1 to optimize the proposed objective functions.

Algorithm 1: Training proposed stain style transfer models with Minibatch Stochastic Gradient Decent.

Input: data $\{X_A, Y_A\}$

Initialized the weights of networks $C(X; \theta_C)$, $G(X; \theta_G)$, $D(X; \theta_D)$

Training $C(X; \theta_C)$ to classify images w.r.t. $\{X_A, Y_A\}$

for number of training iterations **do**

Sample minibatch of 2m images, half normal images and the other half tumor images: $x_A^1, x_A^2, x_A^3, \ldots x_A^{2m}$;

Feed images to $G(X; \theta_G)$ and generate $\dot{x}_A^1, \dot{x}_A^2, \dot{x}_A^3, \ldots \dot{x}_A^{2m}$;

Update $D(X; \theta_D)$ by ascending its stochastic gradient:

$\nabla_{\theta_D} \frac{1}{2m} \sum_{i=1}^{2m} \left(\log D \left(x_A^i; \theta_D \right) + \log \left(1 - D \left(\dot{x}_A^i; \theta_D \right) \right) \right)$;

Feed original images $x_A^1, x_A^2, x_A^3, \ldots x_A^{2m}$ and generated images $\dot{x}_A^1, \dot{x}_A^2, \dot{x}_A^3, \ldots \dot{x}_A^i$ to $C(X; \theta_C)$ to obtain feature representations $F(x_A^i)$ and $F(\dot{x}_A^i)$;

Update $G(X; \theta_G)$ by descending its stochastic gradient:

$\nabla_{\theta_G} \frac{1}{2m} \sum_{i=1}^{2m} \left(\log \left(1 - D \left(\dot{x}_A^i; \theta_D \right) \right) + l_{\text{reco}}(x_A^i, \dot{x}_A^i) + \text{KL} \left[F \left(x_A^i \right) || F \left(\dot{x}_A^i \right) \right] \right)$

end for

Output: Networks model $G(X; \theta_G)$, $D(X; \theta_D)$

---

3   Experimentation

Dataset: We conduct the experiment based on the Camelyon16 dataset [22], which is composed of 400 WSIs from two different institutes, Radbound and Utrecht. The WSIs in these two institutes demonstrate same type of tissue with different stain appearance due to variance in slide preparation. In this study, 100,000 256×256 patches are randomly extracted from WSIs generated in Radbound institute for training, and another 20,000 256×256 patches from Radbound are used for validation. The test set contains 80,000 256×256 patches randomly extracted from Utrecht WSIs. Tumor patches are extracted from tumor regions in tumor slides and are treated as positive samples. Normal patches are extracted from non-tumor and non-background regions in normal slides and are treated as negative samples. The number of positive and negative patches are the same in all the training, validation and testing set.

Experimental Setup: To start with, we use SGD to optimize the diagnosis net with a learning rate of 10^{-3}, batch size of 8 on training set for 100 epochs. Then we use SGD with a learning rate of 10^{-4} and a batch size of 4 to train the GAN based style transfer model on the training set for 60 epochs. Hyper-parameters (i.e. weights of different loss functions) are turned using the validation set. Briefly, different hyper-parameters are tried and the set that generates the highest classification accuracy on the validate set is selected and used in subsequent comparison evaluations.

To evaluate the proposed method, extensive experiments are performed[1]. First, we execute style transfer on test images from Utrecht institute and examine generated images qualitatively in terms of stain resemblance to template images, color consistency, and preservation of histological information. Second, we evaluate the learning efficiency of our model and investigate the effectiveness of the SSIM and DSCSI based loss function in GAN learning. To this end, we record the image reconstruction loss in training to trace the optimization procedure.

For comparison, we also apply above experiments to prior arts by Cho [9], Zanjani [23], Janowczyk [4], Li [24] and Macenko [25]. Among these methods, Cho's method and the proposed methods have similar architectures. Based on GAN, Cho's method generates normalized images from the grayscale versions of query color images and uses MSE as reconstruction loss. Cho and Zanjani both took advantage of deep learning to generate color-normalized images. Note that compared to deep learning methods which learn stain and pattern features in the whole dataset, the last three methods use one slide as reference. For fair comparison, for the last three non-deep learning methods, we randomly choose patches from the training set as reference and execute color normalization over test set. We repeat the experiment ten times and report the average results.

Results and Discussions: First, we compare different normalization methods by visual examination[2]. As you can see in Fig. 2, the target image sampled from training set serves as a comparison reference and source images are sampled from the test set. Comparing all the generated images of three samples, we observe that the first four methods obtain better results than the last three non-deep learning methods. Among the four deep learning methods, images generated by the DSCSI-GAN and SSIM-GAN methods have comparatively more consistent stain appearance and the former has clearer histopathological structure. For instance, for Cho's result, the pink tissue marked by yellow rectangular in Source 1 disappears and tissues marked by yellow circular have stains inconsistent with

---

[1] Due to page limitation, we present two experiments in this paper. More experimental results are provided in the Supplementary file.

[2] Please refer to the file for more color-normalized images generated by various methods.

the sources. The disadvantage in Cho's method is attributed to the discard of color information in stain transfer and the misuse of mean-squared-error(MSE) for image reconstruction. Images generated by Zanjani's method seem better than those from Cho's method and have consistent stain style as well as darker stain in cell areas.

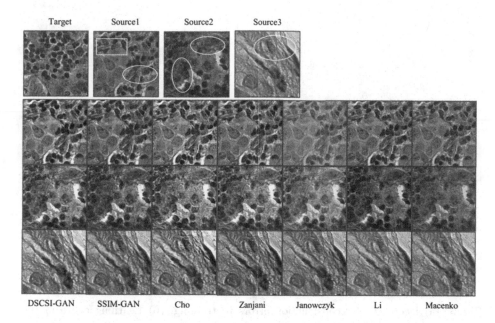

Fig. 2. Stain normalization results of source images by different normalization methods. Target image in first row is sampled from training set as reference image and Source images 1–3 are sampled from testing set for visualization. Images from row 2–4 respectively show the outputs of source images 1,2,3 generated by different methods. Images generated by last three non-deep learning methods appear quite different from target image. In the source images, circular and rectangular markers respectively show the areas that have problems of color inconsistency and histological information loss. The images generated by DSCSI-GAN and SSIM-GAN have comparatively more consistent stain appearance and the former ones have more clear histological structure. (Color figure online)

Second, we investigate the effectiveness of the proposed structure-preserving reconstruction loss on GAN's learning efficiency. Under the same experimental setting, we record the learning efficiency by SSIM based error, DSCSI based error and MSE in Cho's method, shown in Fig. 3(a). Compared with Cho's method where MSE value fluctuates severely, SSIM-GAN and DSCSI-GAN converge more quickly with steady decline. We also use diagnosis network to classify tumor and normal images and compute AUC, Precision, Recall and Accuracy score during training. The changes of four scores are shown in Fig. 3(b). The AUC value, Recall and Accuracy increase more quickly and reach highest values in SSIM-GAN than others. Cho's model is the slowest among the three. These results demonstrate higher efficiency and better optimization performance of the proposed structure preserving loss in this study.

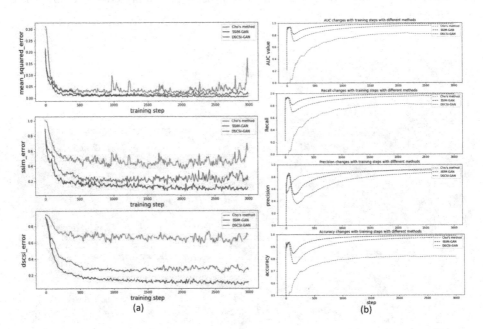

Fig. 3. (a) Decrease of reconstruction errors in training. (b) Learning curve in the context of histopathology image diagnosis.

In summary, our methods have two advantages. First, opposed to traditional stain normalization methods, our model learn color distribution referencing to the whole training set not a single template. This helps to reduce the sensitivity of the normalization method to a particular training case. Second, we propose the use of structural similarity metrics, SSIM and DSCSI, as measurements of reconstruction error in GAN's training and induce the generator to generate high quality images. Since the introduced metrics drive the generator to learn structural content in the hue, chroma, lightness domains effectively, the learning efficiency is high in training.

4   Conclusion

This work presented two stain style transfer models to solve the stain variation problem in computational histopathology. We took advantage of GAN that could learn stain distributions from a template dataset and obtain strong generalization capability to transfer the stain pattern to other datasets. In the proposed methods, we exploited SSIM and DSCSI to construct the reconstruction loss which prompt the model to maintain texture, structure, and color features in original images. Extensive experimentation on publicly available dataset demonstrates that the proposed framework outperforms prior stain normalization solutions in generating stain-consistent images, preserving histopathological information, and obtaining high training efficiency.

References

1. Lyon, H.O., et al.: Standardization of reagents and methods used in cytological and histological practice with emphasis on dyes, stains and chromogenic reagents. Histochem. J. **26**, 533–544 (1994). https://doi.org/10.1007/BF00158587
2. Reinhard, E., Ashikhmin, M., Gooch, B., Shirley, P.: Color transfer between images. IEEE Comput. Graph. Appl. **21**, 34–41 (2001)
3. Tabesh, A., et al.: Multifeature prostate cancer diagnosis and Gleason grading of histological images. IEEE Trans. Med. Imaging **26**, 1366–1378 (2007)
4. Janowczyk, A., Basavanhally, A., Madabhushi, A.: Stain normalization using sparse autoencoders (StaNoSA): application to digital pathology. Comput. Med. Imaging Graph. Off. J. Comput. Med. Imaging Soc. **57**, 50–61 (2017)
5. Khan, A.M., Rajpoot, N.M., Treanor, D., Magee, D.R.: A nonlinear mapping approach to stain normalization in digital histopathology images using image-specific color deconvolution. IEEE Trans. Biomed. Eng. **61**, 1729–1738 (2014)
6. Vahadane, A., et al.: Structure-preserving color normalization and sparse stain separation for histological images. IEEE Trans. Med. Imaging **35**, 1962–1971 (2016)
7. Onder, D., Zengin, S., SarIoğlu, S.: A review on color normalization and color deconvolution methods in histopathology. Appl. Immunohistochem. Mol. Morphol. AIMM **22**(10), 713–719 (2014)
8. Bayramoglu, N., Kaakinen, M., Eklund, L., Heikkilä, J.: Towards virtual H&E staining of hyperspectral lung histology images using conditional generative adversarial networks. In: 2017 IEEE International Conference on Computer Vision Workshops (ICCVW), pp. 64–71 (2017)
9. Cho, H., Lim, S., Choi, G., Min, H.: Neural stain-style transfer learning using GAN for histopathological images, CoRR, vol. abs/1710.08543 (2017)
10. Lahiani, A., Navab, N., Albarqouni, S., Klaiman, E.: Perceptual embedding consistency for seamless reconstruction of tilewise style transfer. In: Shen, D., et al. (eds.) MICCAI 2019. LNCS, vol. 11764, pp. 568–576. Springer, Cham (2019). https://doi.org/10.1007/978-3-030-32239-7_63
11. Shaban, M.T., Baur, C., Navab, N., Albarqouni, S.: Staingan: stain style transfer for digital histological images. In: ISBI (2019)
12. Cai, S., et al.: Stain style transfer using transitive adversarial networks. In: Knoll, F., Maier, A., Rueckert, D., Ye, J.C. (eds.) MLMIR 2019. LNCS, vol. 11905, pp. 163–172. Springer, Cham (2019). https://doi.org/10.1007/978-3-030-33843-5_15
13. Ruifrok, A., Johnston, D.: Quantification of histochemical staining by color deconvolution. Anal. Quant. Cytol. Histol. **23**, 291–299 (2001)
14. Goodfellow, I.J.: Generative adversarial nets. In: NIPS (2014)
15. Wang, Z., Bovik, A.C.: A universal image quality index. IEEE Signal Process. Lett. **9**, 81–84 (2002)
16. Lee, D., Plataniotis, K.N.: Towards a full-reference quality assessment for color images using directional statistics. IEEE Trans. Image Process. **24**(11), 3950–3965 (2015)

17. Ronneberger, O., Fischer, P., Brox, T.: U-Net: convolutional networks for biomedical image segmentation. In: Navab, N., Hornegger, J., Wells, W.M., Frangi, A.F. (eds.) MICCAI 2015. LNCS, vol. 9351, pp. 234–241. Springer, Cham (2015). https://doi.org/10.1007/978-3-319-24574-4_28

18. He, K., Zhang, X., Ren, S., Sun, J.: Deep residual learning for image recognition. In: 2016 IEEE Conference on Computer Vision and Pattern Recognition (CVPR), pp. 770–778 (2016)

19. Wang, Z., Bovik, A.C., Sheikh, H.R., Simoncelli, E.P.: Image quality assessment: from error visibility to structural similarity. IEEE Trans. Imaging Process. **13**, 600–612 (2004)

20. Zhao, H., Gallo, O., Frosio, I., Kautz, J.: Loss functions for image restoration with neural networks. IEEE Trans. Comput. Imaging **3**(1), 47–57 (2017)

21. Zhang, X., Wandell, B.A.: A spatial extension of CIELAB for digital color-image reproduction. J. Soc. Inf. Display **5**(1), 61–63 (1997)

22. Camelyon (2016.) https://camelyon16.grand-challenge.org

23. Zanjani, F.G., Zinger, S., Bejnordi, B.E., van der Laak, J.: Histopathology stain-color normalization using deep generative models (2018)

24. Li, X., Plataniotis, K.N.: A complete color normalization approach to histopathology images using color cues computed from saturation-weighted statistics. IEEE Trans. Biomed. Eng. **62**(7), 1862–1873 (2015)

25. Macenko, M.: A method for normalizing histology slides for quantitative analysis. In: 2009 IEEE International Symposium on Biomedical Imaging: From Nano to Macro, pp. 1107–1110 (2009)

Author Index

Printed in the United States
By Bookmasters